Great Comets

Spectacular and mysterious objects that come and go in the night sky, comets have dwelt in our popular culture for untold ages. As remnants from the formation of the solar system, they are objects of key scientific research and space missions. As one of nature's most potent and dramatic dangers, they pose a threat to our safety — and yet they surely contributed to the origin of life itself.

This beautifully illustrated book tells the story of the biggest and most awe-inspiring of all comets: those that have earned the title "Great." It focuses on Great Comets Hyakutake in 1996 and Hale-Bopp in 1997, which gripped attention worldwide because, for many, they were the first comets ever seen. For everyone interested in astronomy, this exciting book reveals the secrets of the Great Comets and provides essential tools for keeping up-to-date with comet discoveries in the future.

ROBERT BURNHAM has been an amateur astronomer since the mid-1950s. He was an editor for *Astronomy* magazine for many years, and editor in chief from 1992 to 1996. He is now a full-time writer, living in Wisconsin, and is the author of many astronomy books, including *Comet Hale-Bopp: Find and Enjoy the Great Comet* (Cambridge University Press, 1997). His e-mail address is rburnham@execpc.com.

COMET HALE-BOPP (BILL WHIDDON)

Robert Burnham

Great Comets

Foreword by David H. Levy

CAMBRIDGE
UNIVERSITY PRESS

PUBLISHED BY THE PRESS SYNDICATE OF THE UNIVERSITY OF CAMBRIDGE
The Pitt Building, Trumpington Street, Cambridge, United Kingdom

CAMBRIDGE UNIVERSITY PRESS
The Edinburgh Building, Cambridge CB2 2RU, UK http://www.cup.cam.ac.uk
40 West 20th Street, New York, NY 10011–4211, USA http://www.cup.org
10 Stamford Road, Oakleigh, Melbourne 3166, Australia
Ruiz de Alarcón 13, 28014 Madrid, Spain

First published in 2000

Printed in the United Kingdom at the University Press, Cambridge

Typeface FF Quadraat 10.75/15pt. *System* QuarkXPress™ [SW]

A catalogue record for this book is available from the British Library

Library of Congress Cataloguing in Publication data

Burnham, Robert, 1947–
Great Comets / Robert Burnham.
p. cm.
Includes index.
ISBN 0 521 64600 6 (pbk.)
1. Comets. I. Title.
QB721.B79 2000
523.6–dc21 98–50546 CIP

ISBN 0 521 64600 6 paperback

Contents

FOREWORD BY DAVID H. LEVY [VII]

INTRODUCTION [1]

1 GREAT COMETS AND ASTRONOMY [5]

2 WHAT MAKES A COMET "GREAT"? [51]

3 GREAT COMET HYAKUTAKE (1996) [77]

4 GREAT COMET HALE-BOPP (1997) [100]

5 SPACE MISSIONS TO COMETS [136]

6 COMETS AND CULTURES [164]

7 DANGER FROM THE SKY [195]

8 STAYING CURRENT WITH COMETS [218]

INDEX [225]

To Patricia, with all thanks and much love

Foreword

David H. Levy

Comets are made of dirt, several kinds of ice, and a lot of passion. I learned that not long after I became interested in astronomy at age 12; there were planets, stars, and galaxies to be looked at, but there were also these mysterious visitors with tails that would appear from time to time — either completely unexpectedly, or by prediction, like clockwork.

Five years later, comets were a consuming passion for me. What brought me around was the passage of a comet — one of the unexpected kind — in the fall of 1965. It had been found by two Japanese amateur astronomers, Ikeya and Seki, who used small, backyard telescopes to catch their cometary prey. I was amazed that these two had the perseverance to watch the sky, night after night, in pursuit of comets. This find was a lucky one, for Comet Ikeya-Seki was headed straight for the Sun. In October 1965 the comet did an acrobatic hairpin turn around the Sun, and a few nights later I saw its magnificent tail arcing out of the St. Lawrence River east of my home.

On December 17, 1965, I began my own search for comets and the exploding stars we call novae. I code-named the program CN-3, and had three aims:

1. to become familiar with the sky through searching for comets and novae,
2. to discover either a comet or a nova,
3. to learn as much as possible about comets and novae through a research program.

For me that night, aims 1 and 3 seemed easy. As I completed high school and went on to university, I passed a lot of pleasant hours under the stars searching for comets, and more hours in the library learning what I could about comets. Frankly, I never expected that aim 2 would ever come true. There are so few comets, and so many people searching for them!

Nineteen years later, aim 2 finally did come true when I discovered Comet Levy-Rudenko after more than 917 hours with my eye at the eyepiece of a telescope. After that I rather got the hang of how to do it. I found two more comets in 1987, another in 1988, and one in 1989. In the summer of 1990 Comet Levy, my sixth discovery, painted a beautiful picture as it seemed to travel down the summer Milky Way. That year the research part of my program uncovered an old nova, now known as TV Corvi, or Tombaugh's Star.

Comet finds continued for me, both visually through backyard telescopes, and photographically as part of the team of Gene and Carolyn Shoemaker. But none of this could have prepared us for what happened on the night of March 23, 1993, when the Shoemakers and I took two photographs of a region of sky in Virgo, a region that included the planet Jupiter. Comet Shoemaker-Levy 9 turned up on our photographs, and it had just been torn asunder by a close approach to the planet Jupiter. When the comet collided with Jupiter 16 months later, it produced the most spectacular explosions ever seen in the solar system. This comet was not famous for what it was, but for what it did. By immolating itself in the clouds of Jupiter, it provided humanity with a lesson on what happens when comets strike planets. Events like this have happened many times in Earth's past, and it was instructive to watch Nature's demonstration of how a comet impact takes place.

As you begin reading this superb book, you should realize that the solar system has about a thousand known comets. Few of these comets have become bright enough or interesting enough to be called "great" — but there is a good chance that, if you haven't seen a great comet yet, you will sometime in the future: a really nice one should appear every decade or two on average. To my mind, though, every comet is great in its own way, including the several comets that carry the name Burnham through the solar system. Although the Robert Burnham who found these comets is not related to the author of this book, I like to think that, as these comets talk to each other late at night, they consider themselves also a tribute to the younger Burnham who wrote the clear and interesting pages of this book.

You might also consider that, of the solar system's thousand comets, one no longer exists. After crashing into Jupiter at velocities in excess of 130,000 miles per hour, the fragments of Comet Shoemaker-Levy 9 have passed into history, just like earlier, long-gone comets that crashed into the primordial Earth, carrying with them the building blocks of life. Had comets not collided with Earth, life probably would never have started here. In a real sense, when you read this book, you are really learning about your own ancestors.

Introduction

Two bright comets glide across the night sky a year apart and once again these ghostly "long-haired stars" grab the public's attention. The appearance of Comet Hyakutake in March 1996 and Comet Hale-Bopp a year later brought comets back into public consciousness for the first time in a decade — since the visit by Comet Halley in 1986. Was it any surprise to see Hollywood pick up on the theme soon after?

When a comet appears that is bright, eye-catching, or awe-inspiring, people usually call it a Great Comet. To the ancients, of course, all comets were "great" in the sense that they carried an important message from the gods (usually one heavy with doom). Today science has found ways to define what a Great Comet is, but most of us know one when we see it. A Great Comet is one that leaves you standing still and staring into the sky the moment you lay eyes on it. It's a comet that makes you reach for a camera to record it, just as you would for any important event. A Great Comet is one whose image lives in your mind for days and weeks after your first sight of it.

The visual impression given by a Great Comet brings me to a major theme of this book — celebrating the beauty of comets. Both Hyakutake and Hale-Bopp were highly photogenic, and sky photographers all over the world stepped up to take their shots at them. Yet these two excellent comets do not stand alone. Many other Great Comets have visited Earth in the fairly recent past, and these pages are filled with photos of them, too.

A second story that this book aims to tell is about the new knowledge that Great Comets have brought to planetary scientists and, vicariously, to all of mankind as well. For certain periods of astronomy's history, the study of comets lay at the forefront of research, and today comets provide our best look at the raw materials out of which the Sun and solar system were born. Both Hale-Bopp and Hyakutake provided outstanding opportunities for discovery, and the revolution is actually just beginning, thanks to a number of spaceprobe missions to comets that are on the drawing boards or in flight already.

Yet if comets embody both beauty and science, there's also a darker side to the whole picture. A further goal of this book, I hope not inconsistent, is to explore mankind's relationship with comets, which has long been troubled. These gossamer objects of the night are unquestionably beautiful — but they can also be a little disquieting. They appear unexpectedly and trace their courses over all parts of the sky. Despite their lovely appearance, they carry an aura of lawlessness. Comets don't obey the rules that govern the nightly changes in the Moon's appearance, for example, nor do they follow the slow drift of the constellations through the seasons. Comets claim a freedom that has long cast them as the rogues of the cosmos.

Countless chronicles and writings from antiquity onwards portrayed comets as destructive or, at the very least, as laden with bad omens. These attitudes began to fade after the Scientific Revolution of the 1600s, and by the nineteenth century astronomical science had pretty well put to rest popular fears. Thus there's a certain irony in entering the twenty-first century with the realization that comets aren't quite so harmless after all. Scientists now know that comets have struck Earth and the other planets many times in the past and can do so at any time today. We find ourselves in the curious position of discarding a former superstition while keeping a wary eye on Earth's neighborhood at the same time. In the years to come, perhaps we'll look back on Hollywood's 1998 astro-extravaganzas, *Deep Impact* and *Armageddon*, as marking the return of lawless and scary comets to popular culture.

Eight chapters may seem like a lot. But the subject naturally divides into these topics and helping readers keep their bearings isn't the least of an author's jobs. Chapter 1 looks at what scientists have learned about comets and how they behave. It's a long chapter because it has to set the scene for everything that follows. With the new discoveries of recent years — the text reflects the published literature through autumn 1998 — much of what it has to say will be new to many readers. Chapter 2 builds on this knowledge and explores the special factors that can turn an ordinary comet into a great one.

Chapters 3 and 4 follow the stories of Great Comets Hyakutake and Hale-Bopp from their discovery to the struggle to fit the new findings into the larger comet picture. A look into the future comes in Chapter 5, which covers space missions to comets — those that have flown already and the exciting new ones of the years just ahead. These will soon bring us incredible sights and wonders, including samples snatched from comet surfaces and brought back to Earth for laboratory study.

Comet visits don't take place just in the vacuum of space. Every bright comet — and certainly every Great Comet — also scribes a trajectory through popular culture. These "cultural comets" form the subject of Chapter 6. Beginning in antiquity, it examines the mixed relationship (mentioned above) that comets have had with humanity, right down to the Heaven's Gate suicides at the time of Hale-Bopp. Chapter 7 then sets aside Hollywood's sci-fi fantasies to examine the actual risks mankind faces from impacts of both comets and asteroids. (Why look at both? Because the two classes of object pose similar threats, and both are targets of the same searches.)

Finally, Chapter 8 provides a resource guide to help you to explore the background of comet astronomy and to stay up-to-date on the subject using printed sources and the Internet.

Planetary scientists are like the wilderness guides of yesteryear. As their research digs into the mysteries that surround comets, they lead us deeper into the science, the history, and the wonders of the solar family and its far-flung regions where comets rule. In the course of writing this book, I have had many discussions with comet scientists, discoverers, and astrophotographers. For their comments, explanations, and ideas I thank them very much. In particular, I'm grateful to Tom Bopp, Clark Chapman, Alan Hale, David Hughes, Hal Levison, David Levy (who also kindly contributed the foreword), Brian Marsden, David Morrison, Steve Ostro, Alan Stern, Paul Weissman, and Don Yeomans for advice and information. In addition, all the other planetary scientists I have spoken to about comets helped make this book better than it would be otherwise. None of them, however, bears any responsibility for errors that remain.

A different but related debt of gratitude is owed to the hundreds of astrophotographers, amateur and professional, who spent many nights under the stars in the quest to catch these beautiful objects on film and computer disk. If the images in this book recall any of the wonder you felt those chilly mornings and evenings on comet-watch — you can thank the resourceful people who captured these splendid sights so beautifully that we can all share in the delight.

I also wish to thank my local public library, the Hales Corners Library, for providing a helpful and comfortable workplace, including Internet access. Similar thanks go to the Golda Meir Library at the University of Wisconsin–Milwaukee, whose collections have been a valued resource for more than 20 years.

rburnham@execpc.com
Hales Corners, Wisconsin
October 1999

1 Great comets and astronomy

As long as there has been a solar system, comets have wandered among the planets. They have influenced Earth and its inhabitants since the very beginning. In fact some of the water filling Earth's lakes and oceans today came from the impacts of innumerable comets during the first few hundred million years of our planet's history. And comets brought to Earth many of the organic molecules necessary for life, if not life itself.

Comets have also figured prominently in mankind's cosmos, ever since human beings first began to look at the heavens in the endless attempt to comprehend them. No one can say just when and who the first astronomers were — nothing even distantly astronomical has survived from humanity's earliest periods. However, the skywatching impulse is probably as old as our species. After all, birds use star patterns as navigation aids when migrating, and it's silly to think that early humans couldn't have done at least as much.

The oldest astronomical record in existence may be a piece of bone from an Upper Paleolithic site in the Dordogne region of France. It is scribed with what could be the phases of the Moon. Believed by archaeologists to be about 32,000 years old, the bone dates from a time during the last Ice Age that saw a creative explosion of technology and artwork. The Upper Paleolithic is the time when cave paintings were created and when the famous "Venus" statuettes were carved. It's arguable (although impossible to prove) that this was also when early modern humans first began to study the sky consciously.

In any case, it's clear that following the movements of the Sun, Moon, and stars engaged the minds of prehistoric peoples all over the world from a very early date. Countless megalithic monuments, temples, and other structures with celestial alignments attest to a fascination with the sky, even where the exact functions remain mysterious.

The roots of today's astronomy, however, lie in the ancient Middle East, where desert skies afforded easy opportunities for observing the heavens. Moreover the birth of agriculture in this same region about 11,000 years ago gave the inhabitants an urgent reason to discover nature's patterns. Hunters are acutely aware of the passing seasons, and the first calendars probably grew out of their efforts to find game more reliably. But farmers are tied to a specific piece of land. Even more than with hunting societies, agricultural people became dedicated watchers of the sky because their lives literally depend on it. In Mesopotamia, their anxious scrutiny of the skies led to an obsession with cataloguing stars and celestial events of all kinds, including the advent of comets. And because the Mesopotamian cosmology was a malevolent and gloomy one, many celestial signs, especially comets, were taken as bad omens. Therein lies a theme we'll explore more fully in Chapter 6.

But if comets have been watched, and followed — and feared — down the ages, understanding their nature and place in the universe is far more recent.

By the fourth century BC, Greek astronomers and philosophers were observing the sky and speculating about it. Their writings on the nature of comets are the first to have survived to any extent. (Indeed, our word comet derives from the Greek and means "the long-haired star.") While some Greek and (later) Roman philosophers focused on the astrological portent of comets, others were struggling to find natural explanations for them. Aristotle (384–322 BC) thought comets were atmospheric — "dry exhalations" from the Earth that caught fire when they rose to the top of the atmosphere and brushed against the first celestial sphere. But the Roman philosopher and writer Lucius Annaeus Seneca (5 BC?–AD 65) believed comets were celestial in nature because they were not affected by winds. He even suggested that observers in the future should take careful note of comets' movements in hopes of discovering where they came from and what they might be.

When the Roman world fell apart in late antiquity, Christianity soon came to control all intellectual life in its former realm. The Church abandoned the progress that had been made with philosophers such as Seneca and reverted to a largely Aristotelian view of the cosmos. This yielded a neat and tidy picture that could be squared fairly easily with Christian theology. Yet the price of such harmony was that astronomical

Great Comets appear to break the rules that govern celestial motion — which told early civilizations that they were immensely powerful signs from the deities ruling the heavens. (Comet Hyakutake; Jon Jensen and Hugh Salamon; Badger, California; 3 minutes on Fuji ISO 400 film; March 24, 1996)

learning stopped dead. With new discoveries stalled even among the learned, popular views of comets remained pretty much what they had been for thousands of years. Chronicles described comets in terms of portents and signs from an angry God, and lurid tales of doomsday spread in the wake of every Great Comet's apparition. And why not, since every theologian and philosopher concurred? This doleful view was not exclusive to Europe: Chinese skylore interwove conventional astronomy with astrology and divination, and it too echoed the tradition of comets-are-bad-omens, making little effort to find out what comets might actually be.

A major shift in astronomy began in the European Renaissance, after the ideas of Nicolaus Copernicus (1473–1543) were published and began to spread. Reviving and modifying an ancient Greek idea, Copernicus placed the Sun at the center of the solar system instead of the Earth, as Aristotle and many others had proposed. Copernicus' model could be faulted in many details, which meant that it failed to gain immediate acceptance, but it did point astronomy in a fruitful direction.

In 1577, Tycho Brahe (1546–1601) studied a Great Comet that appeared in November that year. Its tail stretched 60° and its head was as bright as the Moon. His careful observations of its position proved that the comet lay at least four times more distant than the Moon, thus proving that comets were not features of Earth's atmosphere. Tycho also noted that the comet's tail always pointed away from the Sun, which suggested to him that the comet was in orbit around it.

Tycho's observational measurements staked out a real advance over ancient knowledge and by implication they also shattered two Aristotelian ideas — that planets travel on crystalline spheres and that comets are atmospheric phenomena. Yet Tycho's finding didn't pass unchallenged. In writings occasioned by three bright comets that appeared in 1618 — the first to be observed by the newly invented telescope, incidentally — Galileo Galilei (1564–1642) dismissed Tycho's observations. He claimed that comets were things within Earth's atmosphere like auroras or sun-dogs. Or, he said, they might be bits of matter ejected from Earth and seen refracted by the atmosphere. One of Galileo's contemporaries, and a fellow-Copernican, Johann Kepler (1571–1630), placed more trust in Tycho's observations, mainly because he had worked with them at considerable length. Kepler noted that the tails of comets always point away from the Sun and argued that this was a clue that they are celestial. Furthermore, Kepler suggested that comet tails exist because the Sun's rays drive off some substance from a comet's head as they pass it. As we'll see, this was a clever guess that hits close to current ideas.

Kepler's work refined the Copernican model of the solar system and gave it mathematical support to parallel the telescopic observations of Galileo. Following Kepler and Galileo, the Copernican universe won increasing converts among the learned — at least in lands outside the grip of the Inquisition — and comet studies began to take on a more modern shape. Astronomers carefully measured comet positions and compiled precise star catalogues and celestial maps by which to gauge cometary motions. Kepler established that planetary orbits were ellipses, although he wasn't sure about comets, whose movements still held many puzzles. Did they move in straight lines? Or curved ones? And if curved, what sort of curve? During the seventeenth century, these ideas were fought over bitterly as astronomers groped toward the concept of universal gravity and how it might control the movements of all celestial bodies in the solar system.

The disputes were finally placed in a mathematical framework by the work of Isaac Newton (1642–1727). His great work, the *Principia* (published in 1687), gave a detailed explanation for the movements of the Moon, planets, and comets. It also explained the behavior of matter in everyday life and showed how this related to the cosmic realm. The *Principia* is arguably the greatest scientific publication of all time. It brought astronomy out of the mists of the Middle Ages and created the foundation of modern physical science.

Newton studied the Great Comet of 1680 using his own observations and those of others, and calculated an orbit for it, the first comet to have one. Although the comet's orbit was very elliptical compared with those of the planets, Newton found that the comet adhered to the laws of orbital motion that Kepler and Newton himself had devised. A number of years later, Edmond Halley (1656–1742) made another important discovery. When he examined the orbital elements of two dozen bright comets, he was struck by the similarity shown by a few. (Orbital

Astronomy (and astrology) grew out of the need for agricultural and hunting peoples to learn the ways of nature and perhaps to control them. Comets were important, if puzzling, parts of the universe. Stonehenge, England's famed megalithic monument, is one of the few structures on Earth old enough to have been around when Hale-Bopp (seen here) paid its last visit over 4,000 years ago. (Paul Sutherland; 24 mm lens; 5 minute exposure at f/4 on Scotchchrome 800/3200 processed at ISO 1600; March 28, 1997)

elements are numbers that uniquely define the size, shape, and orientation of an orbit in space.) In particular, Halley noticed that the orbital elements of the Great Comet of 1682 were virtually identical to those of the Great Comets of 1531 and 1607. Guessing shrewdly, he proposed that all three comets were actually the successive appearances of a single comet which kept returning about every 76 years.

Unfortunately, Halley died before the prediction he made was fulfilled, namely that the Great Comet of 1682 would come back again in late 1758 or early 1759. As everyone today knows, he was exactly right about its orbit and the comet now bears his name. The return of Edmond Halley's comet as predicted gave astronomers of the time great satisfaction because it confirmed Newton's theories about gravity. It also helped put an end to the supernatural image of comets by showing that they are predictable celestial objects that belong to the Sun, just as planets do.

After the epochal findings of Newton and Halley, astronomers devoted much attention to discovering comets and computing their orbits, and the mathematical work of the period is outstanding. However, these orbital calculations said nothing about what comets *were* and, for simplicity's sake, astronomers treated comets as massless points serenely gliding through empty space. It wasn't until well into the nineteenth century that anyone began to make much headway on the nature of comets as physical bodies. Johann Encke (1791–1865) calculated the orbits of numerous comets and discovered, in the same way Edmond Halley had, that one particular comet seen in 1786, 1795, 1805, and 1819 kept returning. Encke determined that the comet could have a period of revolution around the Sun of about 3.3 years, if he made allowances for missing returns. (These occurred because the comet was too close to the Sun to be seen.) Encke calculated that the comet would next return in May 1822, and he was precisely on target.

Encke's comet was the second return of a comet to be predicted and, as with Halley, the returning comet was soon named for Encke himself. But Encke found a curious oddity about the "new" comet. Something was shortening its period by several hours on each orbit. This finding disquieted him and other astronomers. Was some thin, barely perceptible substance filling interplanetary space? (Its drag would move the comet closer to the Sun, resulting in a slightly faster orbit.) Or, much worse, did some unknown flaw lurk in Newton's mathematics? The first explanation was far more desirable, although no such effects had been seen with the admittedly more massive planets.

In 1836, Friedrich Wilhelm Bessel (1784–1846) hit upon an answer: fountains of gas from the comet's nucleus — telescopically visible in the inner coma — might act like little rockets to change the comet's orbit. The classical view of comets had never envisioned such "secular accelerations" or "nongravitational forces" (the modern terms). But with a single stroke Bessel's inspired guess cleared up the mystery, even though a full physical explanation still lay more than a century in the future. Bessel's finding also dramatically underscored the importance of learning more about the nature of comets.

Around this time other studies began to link comets with showers of shooting stars or meteors. A key event in focusing scientific attention on the question was the meteor "storm" of November 13, 1833. Although it had been noted that many meteors appeared every year around the middle of November, the shower in November 1833 was exceptionally rich. Numerous onlookers compared the meteors to snowflakes in a blizzard. They flew so thickly that a great many people were very frightened, thinking the end of the world was at hand. Those who kept their heads, however, could see that the meteors were coming from a single point in the sky, which is now termed the radiant. (The meteors in a shower actually travel on parallel paths, but perspective causes their tracks to appear to diverge.) Since the radiant lay in the constellation of Leo, the meteors were named the Leonids.

Over the decades following the 1833 event, astronomers concluded that the Leonids, the Perseids of August, and other annually recurring meteor showers were celestial in nature and not atmospheric phenomena like clouds or rain storms. They found that meteor showers were caused when Earth ran into streams of meteoric material traveling around the Sun in orbits that were similar in shape to those of comets. As the particles, which range in size from microscopic grains to ones perhaps as large as pebbles, zip into the atmosphere at many miles per second, they vaporize leaving a bright, quick streak in the sky — the classic "shooting star."

Comets universally foretold disasters and misfortune until the start of the scientific revolution. For example, public fears linked the Great Comet of 1556, shown in a woodcut from a German pamphlet, with earthquakes in that same year. Another century and more would pass before the growth of mathematical astronomy could seriously challenge credulous views of comets. (International Halley Watch, NASA/JPL)

Puzzle pieces began falling into place. Daniel Kirkwood (1814–1895) proposed that meteor streams were rivers of debris shed by old or burned-out comets which had become spread around the comets' orbits. Checking past records, astronomers in the 1850s noted that a strong Leonid display had also occurred in 1799, but the annual Leonids since then were not very impressive, except those for 1833. Working independently, Heinrich Wilhelm Olbers (1758-1840) and Hubert Newton (1830–1896) predicted that the Leonid stream would produce another great storm in 1866, which it did.

By that time, however, observers had found the comet that was the source for the Leonids, Comet Tempel-Tuttle. They knew they had the culprit because the comet and the meteors followed the same orbit. (This bolstered an earlier discovery that linked the Perseid meteors to a different comet, Swift-Tuttle.) And having strong Leonid displays at 33-year intervals comes about because this is the period of revolution

As astronomers began to explore the physical nature of comets, they found that meteors came from dust left by comets. Major meteor showers such as the Leonids, seen here as they appeared on November 12, 1799, occurred as Earth plowed through a river of comet dust. In 1833 and 1866 additional recurrences of a Leonid meteor "storm" helped lead to the discovery of the parent comet from which the meteors came. (From *The Midnight Sky*, by Edwin Dunkin, 1869)

for Comet Tempel-Tuttle. Leonid storms happen when Earth passes through the stream shortly after the comet has passed by and its debris is strewn most thickly.

Nor have the Leonids ceased to roar. The last great Leonid storm was in 1966. At its peak, observers estimated they were seeing about 40 Leonids *every second.* The shower on November 16–17, 1999, may be outstanding as well — and I don't know about you, but I'm planning to be outside that night unless skies are solidly overcast.

For the astronomers of the nineteenth century, the discovery that comets were steadily losing solid material to space gave an important clue to their nature. Additional evidence came from other comets of the period, such as Biela's. After being observed at several returns, Comet Biela amazed astronomers by splitting in two in 1846. It returned as a double comet in 1852, but it was never again seen and astronomers concluded it had evaporated completely. In its place, however, was a new meteor shower: the Andromedids.

As years rolled by, new techniques emerged for comet study. In 1858 the first photograph of a comet (Donati's) was taken, but emulsions were slow and comet photography didn't replace careful drawings made at the eyepiece until the 1890s. In 1864, Giovanni Battista Donati (1826–1873), the man who discovered the Great Comet of 1858, used a spectroscope to observe three broad bands in the spectrum of Comet Tempel. These bands were later identified as being produced by carbon. Spectroscopic observations of other comets found solid particles which reflected sunlight and were taken to be dust — the very debris that created meteor showers when they raced into the atmosphere. Spectral observations found uncommon molecules like CN (known in its laboratory form as cyanogen) that are unstable (and highly poisonous) under normal earthly conditions.

Armed with such observations, astronomers entered the twentieth century with some basic notions of what a comet was. While its tail might reach for millions of miles, a comet's nucleus was small, rich in dust and exotic gases, and fragile. It was commonly supposed to be a kind of loosely aggregated sand or gravel bank, whose grains were coated with ices.

Unfortunately, the promising start of cometary astrophysics sput-

At its most spectacular, Donati's Comet of 1858 made a striking sight that impressed many. It's easy to see how in former times the long, curving dust tail of a comet could be likened to a scimitar — held to mankind's throat, if one were feeling superstitious. (Ruth S. Freitag, Library of Congress)

tered nearly to a halt during the first half of the 1900s. The reason was simple. A new generation of large telescopes and instruments had been created, principally the 100-inch telescope at Mt. Wilson and later the 200-inch on Palomar Mountain. Coupled with the development of relativity and quantum theory in physics, they opened up exciting new fields of research such as stellar astrophysics and cosmology. These attracted many researchers and much more public interest than solar system topics did. Any ambitious astronomer of the 1930s knew better than to mess around with "local" things like comets when one could untangle the fates of stars and galaxies or map the expanding universe.

Modern comet astronomy began in years 1949 to 1951, when four fundamental ideas set the course of comet studies on its present path. First, Fred Whipple proposed a model for comet nuclei that still holds up. Then Ludwig Biermann described how a comet's tail emerges from the interaction of the Sun and comet gases, and Jan Oort explained where new comets came from. Finally, Gerard Kuiper suggested that just outside the zone of the planets lies a reservoir of comets, leftover icy fragments of the cloud of dusty gas from which the solar system formed.

The comet nucleus model developed by Fred Whipple (1906–) was soon dubbed the "dirty snowball," the nickname it still carries. Whipple argued that, instead of being an ice-coated sandbank, a comet nucleus is a big chunk of water ice mixed through and through with a dust that consists of silicate- and carbon-rich mineral grains. As a test, Whipple applied his model to Comet Encke and successfully worked out the physics of how Bessel's idea of jets could actually change the comet's orbit. He could also use the snowball model to explain a similar effect on Comet Halley, which likewise exhibited small changes in its orbit from one apparition to the next.

Over the years, observations have added many details to the basic dirty snowball model. For example, while the ice in the nucleus is mostly frozen water, additional ices are also present, such as carbon monoxide, carbon dioxide, formaldehyde, methanol, plus a host of other chemical species. And dust may account for a third to a half of the total mass of the nucleus.

In size, comet nuclei range from a few dozen yards across to perhaps

When sunlight warms the nucleus of a comet, its ices give off a cloud of gas and dust called the coma, at the heart of which lies the tiny nucleus, too small to be resolved in Earth-bound telescopes. (Comet Hale-Bopp; David Lane, 500 mm f/5 lens, 16 minutes on Fuji HG 400; April 8, 1997)

25 miles (40 km) in diameter, the estimated size of Hale-Bopp. Bigger comets than that probably exist, however, including some unusual objects, whose behavior straddles the borderline between comets and asteroids. Among these are 2060 Chiron, which has a diameter of 90 to 150 miles (150 to 250 km) — see "When is a comet not a comet?," page 74. The nucleus of Comet Halley is so far the only one actually photographed from close range (by the Giotto spacecraft in 1986). It therefore serves as a benchmark. Halley's nucleus is an oblong measuring about 10 miles by 5.2 by 5.1 (16 km by 8.4 by 8.2). (Coincidentally, this is nearly the same size as Phobos, the larger moon of Mars.) Halley's nucleus is rotating in a complex way: around one axis it spins once every 7.1 days, while around another axis it rotates every 3.7 days. These motions probably come from the activity of the jets, which can alter a comet's spin just as they can change its orbit. Comet rotations seem to vary widely, from as little as about 4 hours to more than a week, but the average is less than 1 day.

The surface of Halley's nucleus is rough and covered with a dark dust rich in carbon and silicates. The silicate dust spans a range in size, from finer than the finest talcum powder — as small as the particles in cigarette smoke — to fluffy aggregates a few millimeters across. The carbonaceous material is similar in size, but in composition it is more like the fine dust from ordinary pencil lead. Spacecraft data show that major constituents in comet dust include molecules made of carbon, hydrogen, oxygen, and nitrogen. Scientists call this material CHON, from the first letters of the elements, which occur within the dust in about the same proportions that they have in the human body.

Giotto's analysis of Halley dust showed that the grains varied a lot in composition, some being richest in carbon, others being dominated by oxygen or iron. A small percentage were rich in magnesium silicate. It's likely that such heterogeneity is common among comets, since they never underwent processes that could stir their ingredients together enough to even out all the differences.

The surface of Halley reflects only about 3% of the light falling on it, the same as soot or fresh tar. But unlike the stuff you'd seal a driveway with, the surface texture of a comet nucleus is fluffy and porous, and it is much less rigid than styrofoam. The depth of the dusty surface layer

remains in dispute — from a few inches to perhaps yards — and in any case it probably varies in thickness from place to place.

As Halley's Comet approached the Sun, the surface dust layer warmed to room temperature and higher. If you could touch it without being burned, your fingers would come away as dirty as if you had stuck them into powdered graphite or laser printer toner. (They would smell pretty bad too!) Not far below the dust, however, lies a chill that exceeds any in our earthly experience. In the places where the exposed ice is boiling away to gas, the ground temperature hovers around –100 °F (–73 °C), but the center of a comet is probably as cold as the depth of space where it formed, or roughly –400 °F (–240 °C).

The topography of Halley's nucleus is difficult to discern from the photos, in which the smallest details are still about 100 yards across. But several hills and possible craters or inactive vent holes are visible, and the surface reflects light in ways that show it is rough on all scales right down to microscopic-size particles. Despite all its gaseous activity, Halley's nucleus was easy to photograph with the spacecraft's camera, which was about 400 miles away at closest approach. This says the cloud of dusty gas being given off by the nucleus is not too dense. If you stood on the surface of Comet Halley, the sky would look somewhat hazy but sunlight would shine clearly enough to throw shadows. Other comets may have different surfaces. Radar signals that had been bounced off the nucleus of Great Comet Hyakutake, a small but quite active comet, revealed that it is shrouded in a blizzard of icy particles in all sizes from tiny grains to boulders.

Seen by the Giotto spacecraft from close up, the nucleus of Comet Halley proved to be an irregular oblong as large as a small city. It was as black as soot and about 10 miles long, and it displayed several jets of dusty gas. Radar shows that other comet nuclei may be veiled in blizzards of icy cobbles and boulders. (Copyright 1986, Max-Planck-Institut für Aeronomie, Lindau/ H. Uwe Keller)

The interior structure of a comet is largely guesswork, although several space missions in development should answer some questions (see Chapter 5). It's not clear if the frigid heart of a comet is a monolithic body or a conglomeration of "cometesimals" — small pieces of ice mixed with dust, each essentially a comet-in-miniature. The latter is probably the case, however, since observations comparing comets like Hyakutake, Hale-Bopp, and others show relative differences in the gases each emits, both in kind and quantity. If a comet nucleus were made up of many cometesimals with slightly different compositions, the result would be like spumoni ice cream — and no comet would have exactly the same gas output as any other.

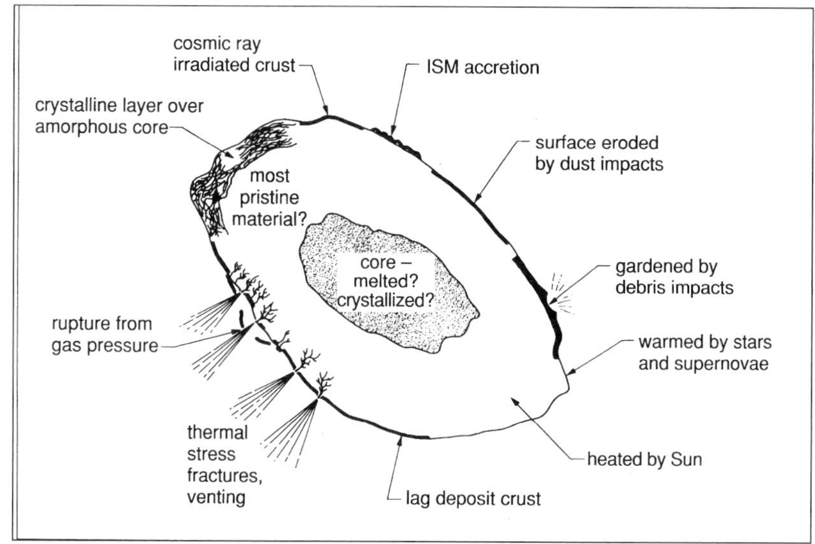

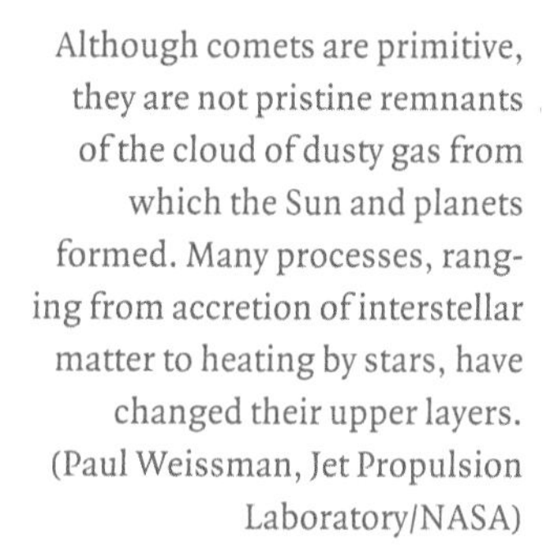
Although comets are primitive, they are not pristine remnants of the cloud of dusty gas from which the Sun and planets formed. Many processes, ranging from accretion of interstellar matter to heating by stars, have changed their upper layers. (Paul Weissman, Jet Propulsion Laboratory/NASA)

More evidence for a conglomerate structure lies in the fact that a comet nucleus is very light for its bulk, with a density averaging less than half that of water. (Think of grocery store white bread.) This suggests that much of a comet's interior consists of microscopic voids, plus interstellar grains coated with particles of ice, all loosely packed or semi-welded together. Many kinds of exotic, even chemically incompatible, ices could be locked in such a matrix. A low-density structure could also explain the fragility seen in some comets, such as Great Comet West, which broke into four pieces in 1976, or Comet Shoemaker-Levy 9, torn into more than 20 pieces by the tidal force of Jupiter's gravity before it crashed into the planet in 1994. Calculations show that a soufflé is more cohesive than the nucleus of Shoemaker-Levy 9 was!

Determining the composition of a comet, however, is a lot easier than weighing it. Celestial objects betray their mass by how they affect other bodies nearby through gravitational pull, yet comets have so little mass that they are always acted upon and never influence another's path. Once again, Halley forms the benchmark although, when the flotilla of spacecraft flew past it in March 1986, its mass was so small that even the closest-flying spacecraft (Giotto, at 400 miles) shot past with its trajectory undeflected. Instead, scientists had to estimate the comet's mass —

As a comet nears the Sun, the gas and dust in its coma form two separate tails. One tail, of ionized gas, shines blue and always points directly away from the Sun. The other tail is of dust, which appears generally white and is usually curved or streaked. (Comet West; Ron Royer and Steve Padilla; March 9, 1976)

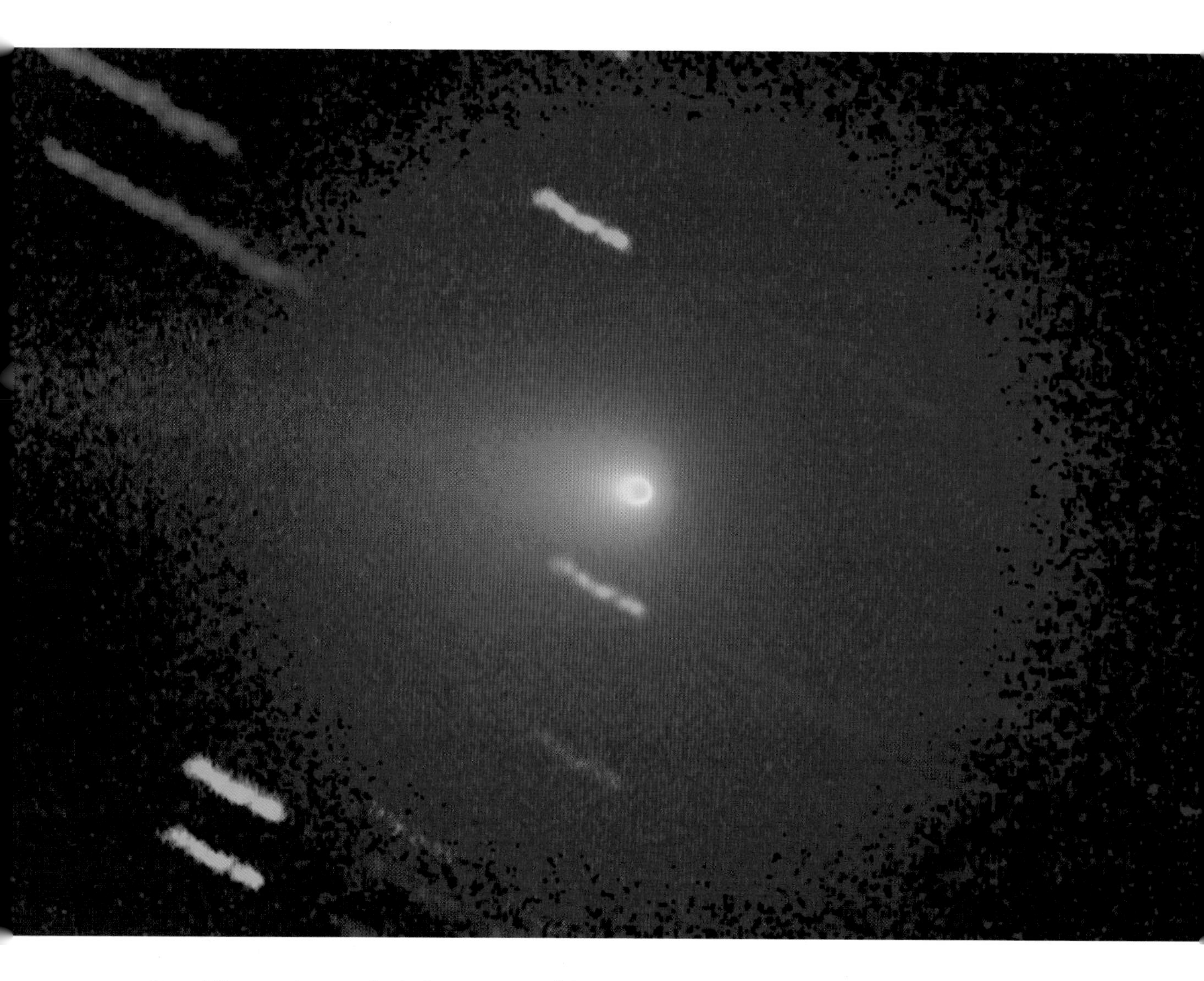

Comet Wirtanen, the target for the Rosetta spacecraft in 2011, is poor in dust. This false-color image from Earth shows the small core of dust around its nucleus as greenish-yellow, the much larger coma of cyanogen in blue and, extending to the left, a reddish gas tail of ionized water molecules. Stars show as colored streaks because the telescope was tracking on the moving comet as it took three exposures through different filters. (Klaus Jockers, Till Credner, Tanyu Bonev, Max-Planck-Institut für Aeronomie)

150 billion (1.5×10^{11}) tons — from the rotation and tumbling movement of the nucleus. Comet Wirtanen (to be visited by the Rosetta mission in 2011) is estimated to have a nucleus about 100 times less massive than Halley's, or 1.7×10^{9} tons. To put these figures into perspective, Halley is as massive as a chunk of solid rock that measures 2.8 miles in diameter (4.4 km) while an all-rock Wirtanen would be much less than a half a mile (well under a kilometer) across.

Boiling away ice each time it comes near the Sun, a comet is constantly losing material. The dimensions of an active comet like Halley or Hale-Bopp shrink by about a meter all around each time it ventures inside the orbit of Mars. Because this loss is never replenished, eventually the comet will die — either by using up all its volatile ices or, more likely, by developing such a thick layer of insulating dust that solar heat cannot penetrate deep enough to activate the ice. The minor planet 3200 Phaethon is probably one such dead comet nucleus. It shows no signs of comet activity, yet its orbit is cometary in shape and it matches that of the Geminid meteor stream. The best explanation is that Phaethon is the nucleus of a comet that became extinct long ago, but not before it had strewn hordes of meteoric dust particles along its orbit. These cause the annual Geminid meteor shower when Earth encounters them every December. How long a comet can remain active depends on many factors. But something like one or two thousand revolutions around the Sun will suffice to shut down a comet's activity, one way or another.

A comet approaching the Sun from deep space — beyond the orbit of Neptune — is an inert body wrapped in a layer of dust. In a telescope it looks like a faint star or asteroid, with no hint of a coma or tail. Solar heat is what turns this frozen hulk into the familiar sight with shining head and long, filmy tail. As a comet approaches the Sun, the growing warmth triggers one kind of ice and then another to flash directly into vapor, with the most volatile ones reacting first. You can see the same process (which is called sublimation) at work on a dry winter day when ice and snow turn directly into water vapor without melting.

Somewhere around the distance of Pluto's orbit (about 40 astronomical units from the Sun), ices made of nitrogen and carbon monoxide pick up enough heat to change to gas. (One astronomical unit is the average distance of Earth's orbit around the Sun. The astronomical unit, abbreviated AU, makes a convenient yardstick for measuring distances within the solar system.) Then, near the distance of Neptune's orbit — 30 AU — methane ice pops into vapor under the weak sunlight, barely 1/900th as strong as at Earth. As the comet comes inside the distance of Saturn, about 9 AU, ammonia ice bursts into gas, adding to the comet's growing "atmosphere." At a distance between Saturn and Jupiter (5.2 AU), frozen carbon dioxide (dry ice) gets going.

However, comets are not made of pure ice, so the activity shown by any given comet may not follow such neat and predictable rules. Complicating factors include the amount of dust that coats the nucleus and how well-intermingled the exotic ices are with the water ice that makes up most of the nucleus. Wherever the dust coating lies thin, the Sun's heat can touch the underlying ice easily. But a thick coating, such as an older comet will collect, tends to blanket the ice so well that solar heat may be effectively blocked until the comet comes quite close to the Sun.

A second complicating factor is the structure of the water ice. Because water ice forms a loose cage of molecules, it tends to trap smaller molecules of the more volatile ices within its fabric where they rest somewhat protected by the surrounding water-ice matrix. This means that some volatile ices may not escape until the protective cage of water ice breaks down and evaporates. This happens around 3 astronomical units from the Sun — about as far as the outer asteroid belt.

Sunlight activates cometary water ice in two distinct stages. When water ice forms at temperatures colder than −215 °F (−137 °C) — commonplace where comets derive from — it is unlike the ice we might drop into a drink or see hanging from the eaves of a house in winter. Scientists term the super-cold form of ice "amorphous" because it lacks the regular latticework of crystals found in the ice we are familiar with. We might think of it as disordered ice. But water ice can remain amorphous only at very low temperatures. If it warms beyond −215 °F, it rearranges itself abruptly and irreversibly into cubic crystals and then hexagonal ones, releasing tiny bursts of energy in the process. The transition causes a chain reaction, as the energy of the change heats the surrounding amorphous ice, which in turn switches to crystalline, releasing its heat, and so on. The reaction halts only when all the amorphous ice in a pocket of the nucleus has converted.

This abrupt change in water ice may help explain the mysterious brightening that a few incoming comets have shown at large distances from the Sun. If pockets of amorphous water ice lie near the surface, a changeover to crystalline ice could power the outbursts. Another possibility is deposits of carbon monoxide or other highly volatile ice at the surface. This material evaporates at very low temperatures and would be one of the first ices to go as a nucleus warms. But not all puzzling comet brightenings can be explained so easily, and some may be due to meteorite impacts on the nucleus.

In any case, by the time a comet nucleus is twice as far from the Sun as Mars (or about 3 AU), a soft fog of gas envelops it. This resembles the haze of vapor swirling around a brick of dry ice at a picnic in July. (Exceptional comets, such as Hale-Bopp, are already quite active by this stage.) The point where the growing warmth causes water ice to start evaporating is sometimes called the "snow line." As the ice evaporates, it creates a cloud of water vapor mixed with the other gases that were formerly locked up in the ice. The gases bursting free of the nucleus also carry off dust particles, generally the smallest ones. Like a deposit of wind-sifted gravel in the desert, the bigger pieces tend to remain behind, slowly building up a layer on the nucleus. Yet comets have so little gravity that even these pieces can be ejected when the gas activity beneath them grows strong enough.

The result of a comet's strengthening activity is the development of a coma, an atmosphere of gas and dusty particles that surrounds the nucleus. Observers with telescopes and binoculars often spot a bright, star-like point in a comet's head. Despite appearances, this is not the real nucleus, which is much too small to be resolved, even using sharp-eyed instruments such as the Hubble Space Telescope. This false nucleus simply marks the most gassy part of the coma, and the tiny, real nucleus lies within it.

The coma grows in size as the comet approaches the Sun and can swell to a diameter of several hundred thousand miles. It grows spherically until the gas becomes so thin that the pressure of sunlight begins to distort its symmetry. The coma usually shows distinct layers or shells in

its inner parts. These occur when the jets on the nucleus turn on and off intermittently as the nucleus rotates. Like the curving sheets of spray shot from a whirling lawn sprinkler, the jetted gas (mixed with loads of entrained dust) drifts outward and forms a shell as it expands into the innermost coma.

Ludwig Biermann (1907–1986) discovered what happens as sunlight interacts with the growing cloud of gas. If water ice and its vapor form the basis for the coma, many of the molecules and atoms in the coma derive from the destructive action of sunlight (mostly at ultraviolet wavelengths) on the water and dust carried off from the nucleus. The main effect of the ultraviolet light is to break so-called parent molecules into fragmentary ones and even individual atoms, and also to strip one or more electrons away from these, leaving them with a positive charge. The resulting gas is called a plasma. Atoms and molecules damaged in this fashion are said to be ionized and they lose their electrical neutrality. This makes the plasma vulnerable to the moving magnetic fields in the solar wind. Ionized atoms are hungry. They are highly prone to capturing stray electrons, thus yielding new molecules as the chemical fragments in the coma jostle and recombine. Some fragments survive only minutes or hours before they rejoin with others to make more stable molecules; others wander for days. The reaction paths are complex and not yet fully mapped. Something like 50 chemical species have been found so far in cometary comas and the inventory is far from complete.

Large as it is, the coma still does not mark the full extent of the comet's head. Surrounding the coma, with about ten times larger a diameter, is a vast halo of neutral hydrogen made of intact atoms driven away from the coma. (Biermann predicted this halo two years before it was confirmed in 1970.)

When Comet Hyakutake passed closest to Earth on March 25, 1996, the Hubble Space Telescope used its high-power camera with a red filter to study the inner coma. Icy regions on the nucleus become active as they rotate into sunlight and eject lots of dust in jets, faintly visible in this image. Sunlight striking the dust eventually turns it around and pushes it tailward (toward the upper left). (H. A. Weaver, Space Telescope Science Institute; NASA)

But the coma is not what makes comets so awe-inspiring — it's their enormous tails. These led the Aztecs to call comets "smoking stars" and the name is remarkably descriptive. The Chinese term for a comet, "broom star," also echoes this appearance. As Johann Kepler guessed, comet tails are born from the coma. They take two easily distinguished forms, one made of dust and the other of ionized gas. Since comets vary in their activity and composition, so does the relative importance of each kind of tail. Dust tails dominate some comets (like Hale-Bopp) while with others (Hyakutake for instance) the gas tail is more prominent.

Yet, for all their impressive length, comet tails are pretty close to nothing, being vastly emptier than the best vacuum achievable in a terrestrial laboratory. Where a comet tail contains only a few thousand atoms and molecules in a volume the size of a sugar cube, the air you are breathing right now packs 30 billion billion atoms and molecules into the same space. The amount of gas in an entire comet tail roughly matches the volume of air inside a sports arena or large theater.

A comet's dust tail consists of the carbon-rich and silicate dust particles blown into the coma by gas escaping from the nucleus. An active comet sheds many tons of dust per second. When the dust particles drift out to reach the outer coma, the pressure of sunlight begins pushing on them and driving them off, with smaller particles being the most affected. Heavier particles respond more sluggishly; they are pushed away also, but more slowly. The dust tail often appears curved, as the nucleus and coma race away from the debris they have shrugged off and the discarded dust particles set out on their own orbital paths in the wake of the comet. Dust tails may stretch for 10 million miles and usually display some structure, often streaks produced by gouts of dust ejected at one time. To the eye the dust tail appears white or ivory, but it frequently looks cream-colored or even yellowish in photographs — the result of its shining by reflected sunlight.

The gas (or ion or plasma) tail appears markedly different. Being made of ionized gases and hence electrically conductive, the material responds to changes in the solar wind. This is a flow of charged particles (electrons and protons) that streams from the Sun at high speed. Near Earth's orbit, the solar wind blows at about 300 miles per second (500 km/s) and, as it strips the ionized gas from the coma, it pulls and tugs on it and shapes it into a long, straight streamer. The ion tail always points directly away from the Sun like a windsock, even when the comet is moving outward from the Sun. Because the solar wind is turbulent, and the comet is often charging across its outward flow, a plasma tail usually shows knots, condensations, and kinks like a long banner flapping on a breezy day. The abrupt changes can snip off portions of the ion tail in what is called a disconnection event (see the photo on page 88), after which a new portion of tail regrows in the space of a few hours. Motions of the ion tail attract much attention from comet and solar scientists because the tail acts as a tracer for the activity of the invisible solar particles, just as waving long grass marks the passage of the wind over a hilltop.

Ion tails differ from dust tails by their color, which is usually a distinct blue or blue-green, both to the eye and the camera. Unlike the dust tails which shine by reflecting sunlight, ion tails shine by fluorescence. In this process the atoms and molecules briefly absorb a bit of energy from sunlight and then re-radiate it. Most of the light in a comet's plasma tail comes from ionized carbon monoxide, nitrogen, and water — although carbon monoxide is the main ingredient in what we see.

Comets are among the solar system members which have changed least since the birth of the Sun and planets. This makes them highly valuable for studying the solar system's earliest period and the ingredients out of which it formed. No one saw the solar system being born, but astronomers can observe other newborn star systems and run computer models that attempt to recreate the scene. Comets yield important data points for those simulations.

In a sense, spiral galaxies like the Milky Way exist solely to manufacture stars (and presumably solar systems), and the galaxy indeed makes a new star every few years on average. It's a process that never quits so long as a galaxy has ample supplies of gas and dust. These may be remnants from the Big Bang or recycled debris that has already spent time inside a star before being thrown back into interstellar space. As planetary scientists now understand it, less than 5 billion years ago, a vast, cold cloud of dust and gas was drifting in the Milky Way galaxy. Could it have done so, a beam of light would have taken a few hundred or a thousand years to drill through the cloud from end to end. In places the cloud was virtually transparent while elsewhere it was more like a thick fog. The cloud rotated slowly relative to the stars around it.

If one of these nearby stars had not exploded as a supernova, the cloud might remain that way still. But the supernova shock wave nudged the cloud so it began to collapse under its own gravity. The explosion also sprinkled the cloud with chemical elements not found among its

original components. The cloud's densest part formed a core that grew larger and hotter as its gravity attracted material. Just as skaters spin faster when they pull in their arms, the infall of debris sped up the cloud's rotation. Meanwhile, the hot core grew massive enough that its central heat and pressure touched the flashpoint for thermonuclear fusion, and the protostar became the Sun. By this time, 100 million years or less had passed since the cloud began collapsing — just moments, astronomically speaking.

A star isn't all that was born from the cloud. The material that didn't fall into the Sun formed a broad disk around it called the solar nebula. It swirled slowly about the infant Sun, like the ingredients for a cake being churned in a mixer. In the nebula, particles of dust collided and stuck together. Lightning from static discharges arced across the disk and through it, melting some grains and vaporizing others. Shock waves rippled through the nebula, and disk particles clumped and grew haphazardly to become planetesimals — small primitive bodies of rock and ice a few miles across. Gradually the hordes of planetesimals accreted into a smaller number of larger bodies, the protoplanets. These in turn merged, often violently, to form the planet-size bodies of today. Near the Sun, the solar nebula was too hot for much water to exist, let alone ice. Planetesimals (and later planets) that formed there had to be made of materials with high melting points, like nickel, iron, and silicate (quartz-rich) rock. From the present orbit of Jupiter outward, however, temperatures fell low enough that bodies could retain significant quantities of water, ice, and other elements easily destroyed by heat. From this crude division arose the solar system's distinct planet populations: small rocky planets like Earth that huddle near the Sun, and larger gas-rich bodies like Jupiter which orbit at a distance.

The trillions of comets that formed from the solar nebula are pieces of it that have always stayed on the sidelines of the action. They are ice-rich for the simple reason that they coalesced at frigid temperatures in the outer solar system, a region that has never felt more than the smallest amount of warmth from the Sun. But comets are not completely untouched and unaltered. Any ices exposed at their surface will be bathed in ultraviolet light from the distant Sun and the surrounding stars of the galaxy. The icy materials will slowly change as they absorb

To the eye, Hale-Bopp's most visible feature was the broad and featureless white dust tail — yet camera and film also caught its blue gas tail clearly. (Loke Kun Tan; 6 × 7 cm format, 165 mm f/2.8 lens, 8 minutes on hypered Kodak Ektar 25 PHR; March 29, 1997)

energy, turning darker and redder over the eons. From time to time, solitary stars will drift near the solar system and gently heat some comets for a few hundred thousand or a million years, perhaps raising the temperature enough to make changes in the more volatile ices. High-speed collisions with interstellar particles may erode their surfaces. Finally, energetic cosmic rays will penetrate comet nuclei and alter their upper layers.

The result is that, over the age of the solar system, comets gradually transform from truly primordial to something more evolved. Even the comets parked for ages in this dim, distant deep-freeze will be altered to depths of 15 to 150 feet (5 to 50 meters). This doesn't make comets any less interesting, because they still are the least-altered objects in the solar system that we know about. However it does mean that scientists have to proceed carefully in their interpretations, especially when they have actual surface samples in the laboratory to analyze, a day not far off.

Planetary scientists draw an arbitrary distinction between comets whose periods of revolution around the Sun last 200 years or longer and those that take less. The former are called long-period comets, the latter short-period. (Making 200 years the dividing line has no physical significance; it was chosen early in the 1900s to reflect the period for which accurate observations existed.) According to this scheme, Halley is a short-period comet thanks to its 76-year orbit, while Hale-Bopp, not due back for 2,400 years, definitely falls in the long-period category.

The division of comets into long period and short period, however, is a man-made notion and it obscures the natural differences that divide comets. Among these are the shapes, sizes, and orientations of their orbits in space. Since these properties stand apart from accidents of human observation, they reflect underlying realities that come from comets' physical natures and their orbital histories.

For example, long-period comets fall into two categories, the first being comets with orbits so large and periods so long that the comet (when seen) is probably making its first trip through the region of the planets. These are often called "fresh" or "dynamically new" or "Oort Cloud" comets, for reasons we'll get to in a moment. The second type, called "returning" comets, contains objects on smaller orbits that must

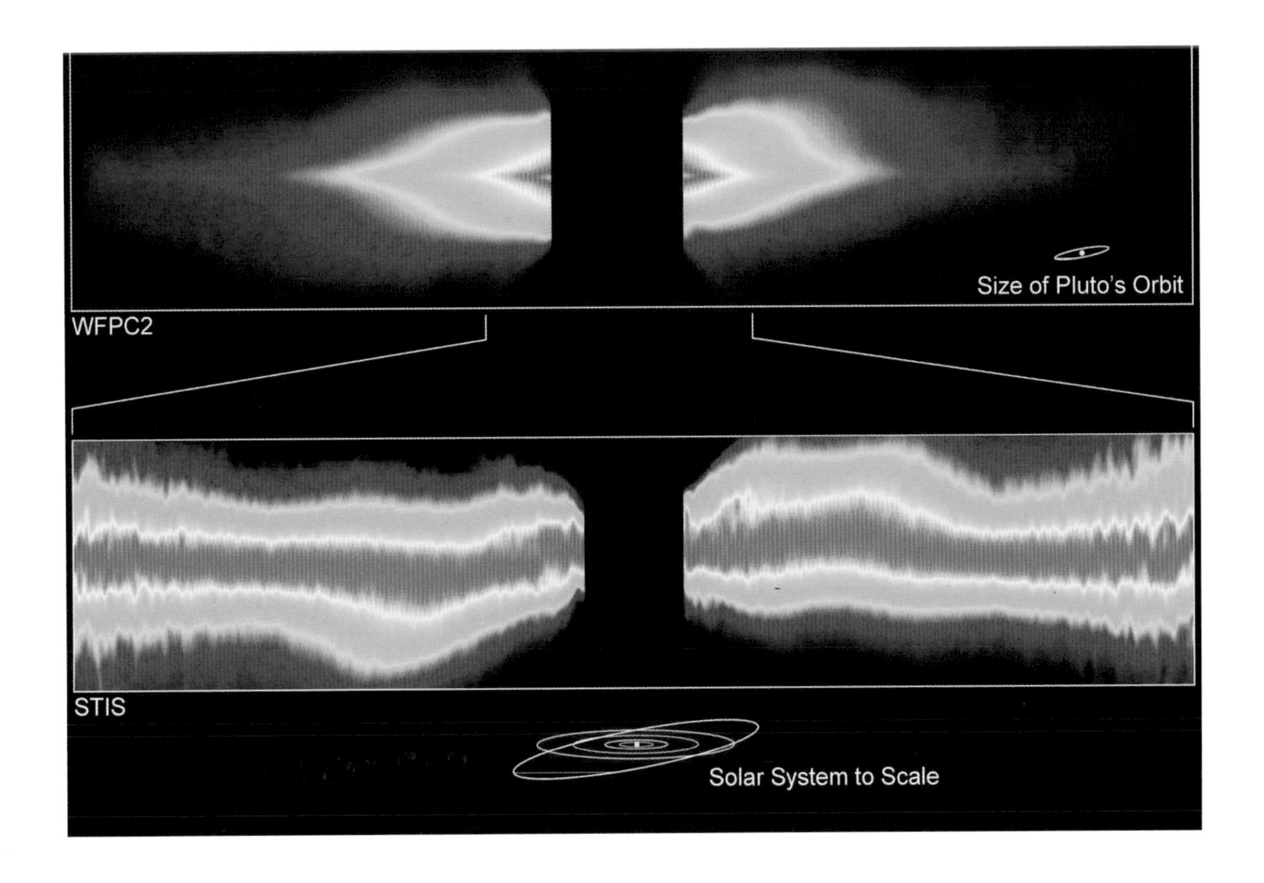

A disk of dust surrounds the star Beta Pictoris, evidence of a solar system in the making. This Hubble Space Telescope view shows the disk's full extent, some 140 billion miles (1,500 astronomical units) wide. A black strip blocks the bright glare of the star. (Space Telescope Science Institute, NASA/Goddard Space Flight Center)

have flown through the inner planetary system several times, even if no reliable record exists of a previous visit. Comet Hale-Bopp fits the latter case, since its last return was some 4,300 years ago. (While this is before any comet observations exist, it's extremely recent in astronomical terms, being less than one-millionth the age of the solar system.)

Short-period comets likewise divide naturally into two groups, the Jupiter-family comets and the Halley-types. As the name suggests, a Jupiter-family comet has its destiny controlled by Jupiter. These comets have a maximum distance from the Sun (a point called aphelion) close to that of Jupiter's orbit, about 5.2 astronomical units (AU). Moreover, the orbit of a Jupiter-family comet inclines to the plane of the solar system (the ecliptic) by less than 45°, and many have inclinations of 15° to 20°. In

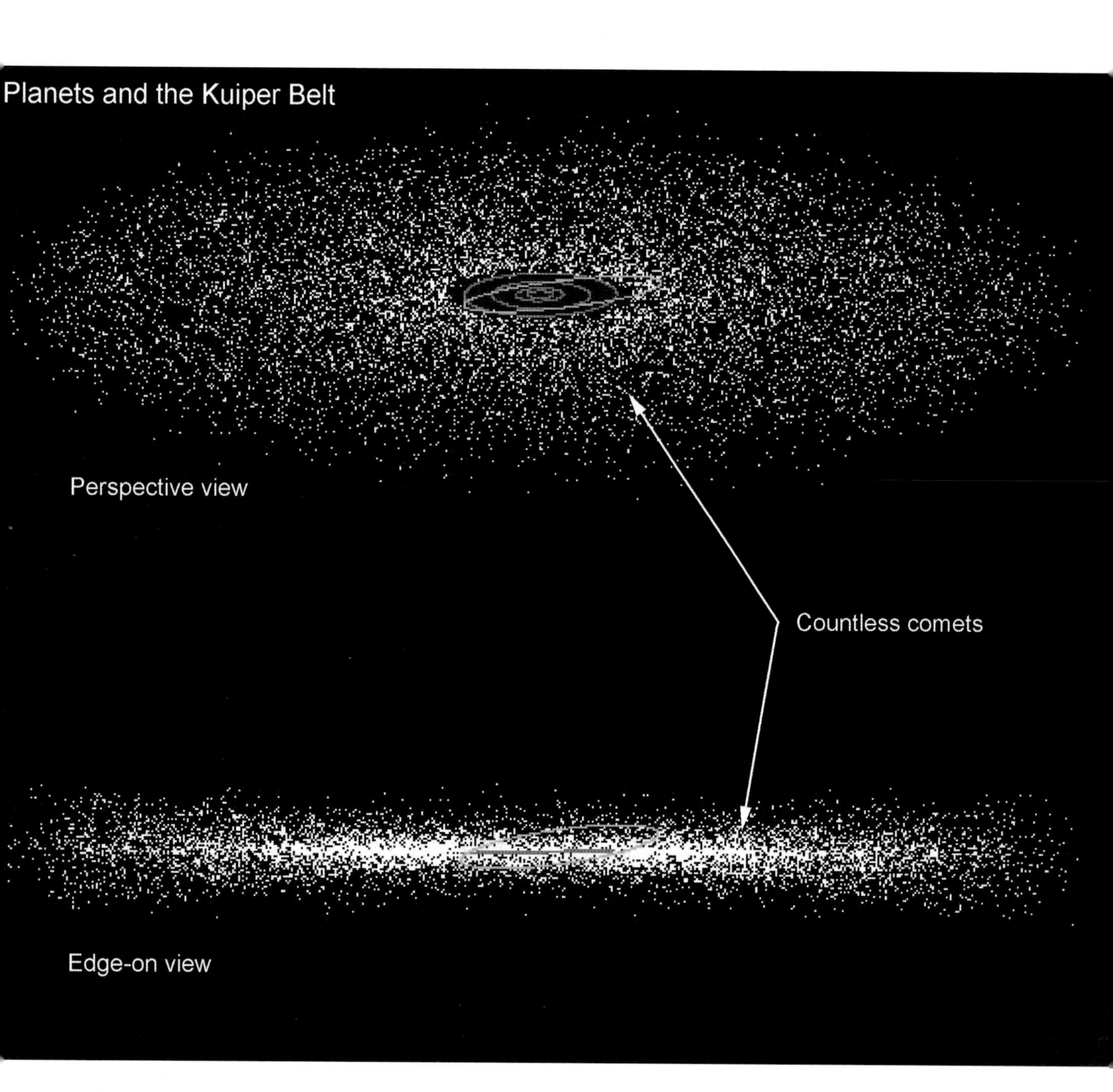

Surrounding the system of planets like a flat collar, the Kuiper Belt forms a reservoir that is the original source for short-period comets like Halley's. The Belt extends from about the orbit of Neptune to many hundreds of times farther from the Sun than Earth; this schematic diagram shows only its innermost part. Planetary orbits from Jupiter to Neptune are in red; the eccentric orbit of Pluto (once a member of the Kuiper Belt) is shown in green. (Harold F. Levison/ Southwest Research Institute, Boulder)

a loose sense, Jupiter-family comets stay fairly close to the plane in which the planets themselves orbit, and they travel in the same direction around the Sun as the planets do.

A Halley-type comet, on the other hand, has an aphelion distance greater than about 7 AU and a period between 20 years and 200 years. Its orbit can incline at any angle, and the direction of the comet's motion around the Sun may run opposite that of the planets. (Comet Halley itself has just such a retrograde orbit.)

To understand how these natural divisions might arise, let's step back to take a look at the big picture of where comets are coming from. In 1950, Jan Oort (1900–1992) published a plot of 19 long-period comets with well-determined orbits. The plot showed a curious fact: these comets' orbital inclinations were oriented at random and over half had aphelion distances of 50,000 to 100,000 AU or more. To put this number in perspective, recall that Pluto, the farthest planet, orbits the Sun at an average distance of only 40 AU. These comets had aphelion distances that lay more than a thousand times farther away. Another way to envision this is to imagine that if we shrink the solar system so that Pluto orbits 1 inch (2.5 cm) away from the Sun, Oort's comets were cruising out to distances of 100 and 200 feet (30 and 60 meters). Comets in such orbits would have periods of revolution around the Sun measured in millions of years.

The random orientation of orbits plus the huge aphelion distances led Oort to propose that the solar system is enveloped in a distant, spherical cloud of comets shaped like a dandelion blossom gone to seed. The idea was eventually adopted and the cloud was named for Oort. In his work, Oort had used a sample of only 19 well-known comets, but present estimates hold that the Oort Cloud in all contains something like 6 trillion individual comets — all adding up to a total mass of roughly 40 times that of Earth.

As currently depicted, the Oort Cloud extends far into space — 50,000 AU is more than three-fourths of a light-year. The cloud extends as far as the Sun can retain a comet against the gravitational tugs of nearby stars and perturbations from massive dust and gas clouds in the Milky Way. Exactly where this boundary lies is undetermined, but a reasonable guess is that the Sun's control peters away within about 2 light-years. For this reason, comets in the inner parts of the Oort Cloud are bound more tightly to the Sun and the outer fringes of the cloud are much less populated. In fact, calculations show a constant loss of comets to interstellar space from the outer edges, a loss which is partly resupplied from the inner portion. The cloud is probably also not perfectly spherical; it would stretch in the direction of the Sun's motion through our arm of the galaxy, with a longer, less-populated tail of comets reaching behind it.

As the Sun travels within the galaxy, the Oort Cloud experiences occasional perturbations from the gravity of passing stars. For example, recent studies have shown that a dim red dwarf star named Gliese 710 will pass through the outer Oort Cloud in about 1.4 million years. In addition, the giant clouds of dusty gas that litter the disk of the galaxy can also exert a gravitational effect on the Sun's cometary retinue. Such galactic perturbations are quasi-periodic in nature, reflecting the Sun's up-and-down motion in the galactic disk during its 250-million-year orbit of the galactic center. Quasi-periodic perturbations will induce comets to leave the Oort Cloud at quasi-periodic intervals. Many perturbed comets will be lost from the Sun's grip, but appreciable numbers must surely fall out of the Oort Cloud into the inner solar system, where they may strike Earth and the other planets.

Some paleontologists report signs that mass extinctions seem to recur every 26 million years; others have reported finding a 30-million-year periodicity for both extinctions and major impact craters. If a periodicity is confirmed, scientists say, then periodic "comet showers" could be the cause. During a comet shower, Earth's skies would probably contain several Great Comets at any given time, and a new one would appear weekly. Under such conditions, collisions with our planet and others would occur much more often than at present — hence the proposed link with extinctions. (Wondering about the effects of Gliese 710's upcoming passage? Calculations show that it should raise the number of comets passing through the inner solar system by no more than about 50%, vastly less than the factor-of-a-hundred or more increases that a full-blown comet shower would probably produce.)

In 1984 one group of researchers hypothesized a "dark star" companion to the Sun to explain periodic extinctions. (Another group proposed a yet-to-be-discovered Planet X orbiting far beyond the known planets.) In a dramatic touch, the distant companion star was named Nemesis.

According to the idea, Nemesis would have a far-flung, highly eccentric orbit that reaches perihelion (its closest approach to the Sun) within the outer Oort Cloud, where its perturbations would send comets down upon our heads. The idea was attention-grabbing and it made headlines, but no evidence has ever been found for Nemesis (or for Planet X) besides the extinction periodicity, if it really exists.

The differing figures for the extinction and cratering periods have cast some doubt on their validity, and many scientists remain uneasy with the statistical sampling methods used to identify the periods. Another problem with Nemesis and Planet X is that impacts from asteroids, not comets, may be the major extinction-inducing agent — and no amount of Oort Cloud perturbations can plausibly drive asteroids out of the Main Belt between Mars and Jupiter and send them on Earth-crossing paths.

More fundamentally, there is considerable doubt whether the Sun could hold on to an object like Nemesis for long at such great distances, given the shifting gravitational tugs from the galaxy. Besides, many astronomers believe that deep all-sky surveys would have by now detected any dark-star companions or large massive objects such as Nemesis. All that can be said at present about periodic extinctions is that the geological record is suggestive, but it needs more research to determine if the period is real and how such a mechanism would actually work.

A comet that will grace our skies leaves the Oort Cloud when it has been perturbed out of its placid stasis. Falling almost imperceptibly at first, it slowly drops through the icy dim twilight toward the bright arc-light of the Sun. After ages pass, it finally races through the planetary system, warms under the Sun's rays, and grows a coma and tail. Then it heads back out again. But whether the comet goes back to its former distance or to some shorter one from which it will return more quickly — or whether it will be kicked out of the solar system completely — all depend on the comet's interactions with the gravity of the planets (mainly Jupiter's) as it flies through the inner solar system.

The Oort Cloud is thus the immediate source for fresh or dynamically new comets and, given a few planetary perturbations, also that of the so-called returning comets. The attrition, however, is brutal. Computer simulations show that the typical Oort Cloud comet makes an average of five revolutions around the Sun, lasting about 600,000 years, before it

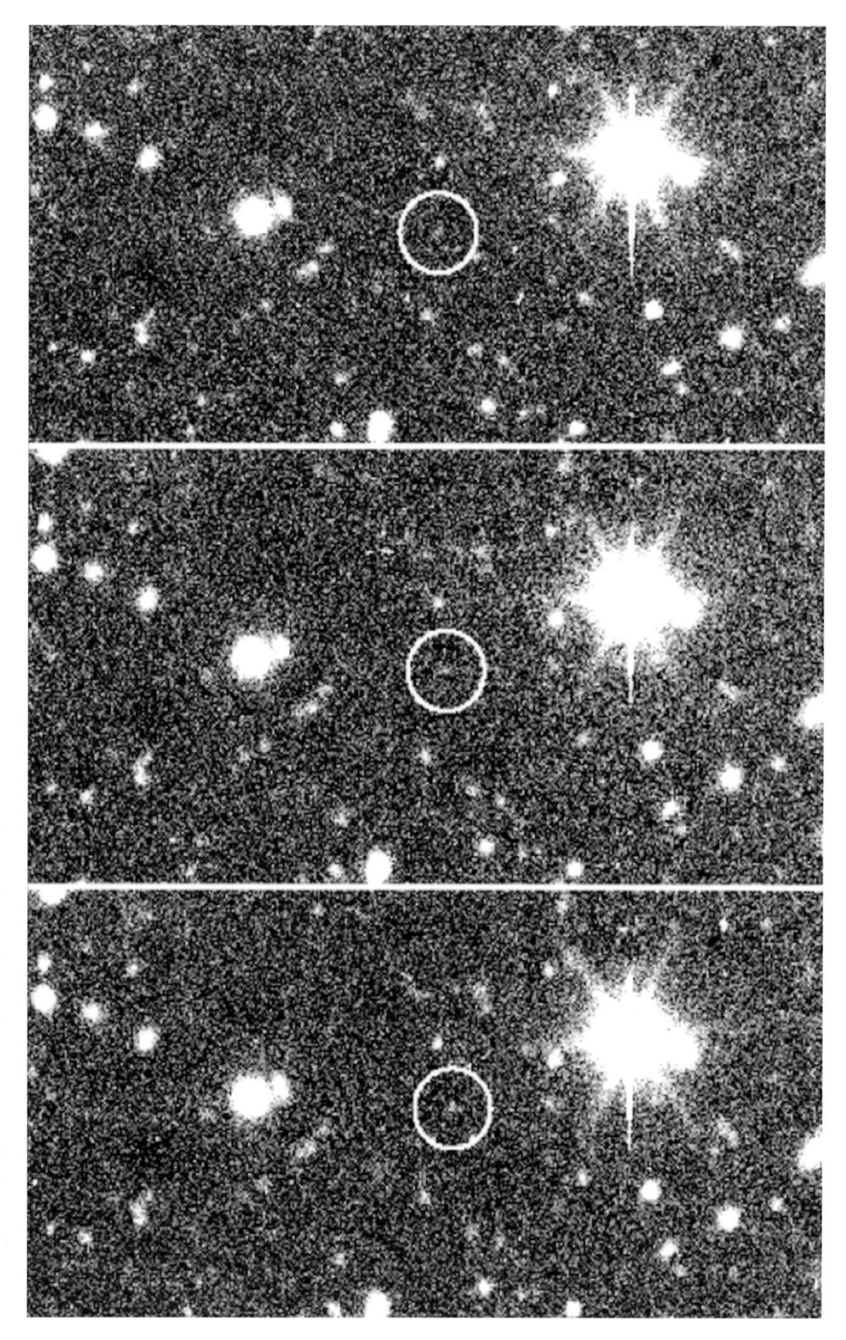

Worlds of twilight. In the icy darkness beyond Pluto, small comets by the billion orbit the Sun in the Kuiper Belt. Tiny and dark, these objects strain the detection limits of even giant instruments such as the 10-meter Keck telescope on Mauna Kea used for these images. The circled object was detected three times, but only on a single night and therefore its orbit is poorly determined. It may never be seen again. (Jane Luu, Harvard University, and David Jewitt, University of Hawaii)

reaches some final state. Only about 4% of the comets falling into the inner solar system from the Oort Cloud return there. Nearly half are ejected by Jupiter on their first pass and ejections reach 64% when subsequent passes are taken into consideration. For another 25%, the end-state is random disruption; they split into small fragments as Biela's Comet did in 1846 and vanish to become part of the haze of minute debris orbiting among the planets. About 7% either boil away all their ices or become inert when their dust mantles grow too thick. Finally, less than 1% shift into short-period orbits or meet their doom by passing too close to the Sun to survive.

Recent work in which scientists create model solar systems inside a computer has led to the conclusion that Jupiter's comet-controlling behavior had a direct hand in keeping Earth habitable. Time and again, the model solar systems show that, without a Jupiter-size object orbiting about where it does, the planets in the inner solar system would get hit by up to a thousand times more comets than they actually did. Life on Earth might have survived if comet impacts happened a little more frequently, but increasing the rate a thousand-fold could well have extinguished all terrestrial life before it really got going. When ancient civilizations identified the planet Jupiter with the king of the gods, they were more right than they ever knew.

While Oort's proposal described where fresh long-period comets were coming from, it explained nothing about where comets originated or formed. For a couple of reasons, it's highly unlikely they were born in the Oort Cloud itself. First the average orbital motion of a particle at that distance is far too slow for mutual collisions to build up comets (or even cometesimals) from the material of the solar nebula, despite having almost 5 billion years to do it. Second, the solar nebula was a flattish disk and objects that formed out of it reflect its flat shape in their own orbits. This is why all the planets — even Pluto — and the Main Belt asteroids orbit relatively close to the plane of Earth's orbit. This broad disk, however, little resembles the spherical cloud of comets that Oort found.

An answer to the origin of comets appeared at the same mid-century point when Oort was proposing the cloud. In 1951 Gerard Kuiper (1905–1973) wondered why the planetary system should seem to end so abruptly at the orbits of Neptune and Pluto. He concluded that beyond these objects there probably lay a multitude of undiscovered smaller bodies, rich in ices condensed from the solar nebula and orbiting close to the plane of the solar system. These would form a wide zone extending outward from about 30 AU, the vicinity of Neptune's orbit. Kenneth Edgeworth (1880–1972) had proposed much the same idea in 1949 but how much Kuiper knew of his work is unclear. These small, remote objects — comets in all but name — could be extremely numerous, Kuiper thought. But they would be orbiting too slowly to have banged into each other often enough to build another large planet like Neptune or Uranus. And so they would remain drifting on the outskirts of the planetary realm down to today — construction debris left behind when the Sun's planet-building contract ran out.

This broad zone is now called the Kuiper Belt, or perhaps more properly, the Edgeworth-Kuiper Belt. It extends from about 30 AU (Neptune's orbit) out to very roughly 1,000 AU. Like the solar nebula it formed from, the Kuiper Belt is relatively flat. At its farthest extent, it may merge with the spherical Oort Cloud. But how and where this merger occurs — if it does — is open to lots of debate because the picture is based almost wholly upon computer models.

Recently, astronomers have begun to detect what may be the Kuiper Belts of other stars. For example, a 4th-magnitude star lying only 10.8 light-years away, Epsilon Eridani, is surrounded by a ring of cold dust (and perhaps comets) that is similar in scale to the probable dimensions of the Sun's Kuiper Belt. Because Epsilon is at an earlier stage of its life than the Sun, astronomers will be studying it closely and comparing it with models of how our own Kuiper Belt grew and evolved. Farther away (about 50 light-years) is 4th-magnitude Beta Pictoris, another young star with a similar disk of dust (see page 35).

In any case, scientists see the inner Kuiper Belt as the original source for the comets in the Oort Cloud and for many of the comets passing near Earth as well, such as the Jupiter-family comets. The chief actor in the drama is Neptune, assisted by Jupiter and the other gas-giant planets. Neptune's gravity can snare bodies in the Kuiper Belt out to 50 AU or farther. It does so by periodically tugging on a Kuiper Belt object and altering the shape of its orbit so that its point of closest approach to the Sun moves inward. This reinforces the tugging because it brings the object closer to where it can be worked upon more vigorously by

Neptune and the other large planets. The eventual outcome of this gravitational come-hither is that many comets will be drawn in from the Kuiper Belt. Then after a close encounter with one of the big planets (usually Jupiter) they will be thrown in every direction on far-sailing courses. Their fates are diverse: perhaps most will go out to create the Oort Cloud, but some will drop sunward where they may fly past Earth. For the others, after a few dozen trips among the planets, their orbits may change into the familiar Halley-type and Jupiter-family patterns — provided the comets don't smash into one of the planets or their moons first.

A small population of large, volatile-rich icy dwarfs called Centaurs now cruises in unstable orbits among the outer planets, like objects caught in transition. The first discovered was 2060 Chiron, 90 to 150 miles (150 to 250 km) in diameter. (It lent its name to the entire group; in Greek mythology Chiron was a centaur.) The Centaurs, ten in number as of mid 1999, are probably giant comet nuclei that arrived from the Kuiper Belt in the last few hundred million years. They may someday become active comets (Chiron already is), but their eventual fates take a familiar course: within a lifetime running from 500,000 years to perhaps 100 million years, they will be ejected from the planetary realm, collide with a planet, or dive into the Sun. But how and when this will happen is unknown because in each case their orbits careen too unpredictably. (See "When is a comet not a comet?," page 74.)

Today's inner Kuiper Belt is believed to hold only a fraction of its original population. After billions of years of winnowing there isn't much left, at least by comparison with what orbited there during the earliest stages of the solar system's formation. Estimates suggest that hardly more than a tenth of an Earth-mass of material still remains between about 30 and 50 AU, down from perhaps 30 Earth-masses at the start. (Much farther out, however, the population should be closer in numbers to its original state.) The giant planets' grab-and-fling activity back then must have been awesome to behold. The skies of early Earth would have been filled with many more comets than we ever see today — and large numbers of them surely struck our planet. These comets delivered some of the water that sloshes through Earth's hydrosphere, and cometary organic chemicals must have contributed much to the origin of life.

The thinning-out of the inner Kuiper Belt was all the more thorough because Uranus and Neptune drifted outward as they formed. Some calculations show that Neptune increased its orbital distance from the Sun by at least 5 astronomical units as it grew to reach its present size, while Uranus moved out somewhat less. (Jupiter and perhaps Saturn moved inward 1 or 2 AU.) As they migrated, the giant planets took their gravitational perturbations along with them. In the case of Uranus and Neptune these swept outward through a large volume of the inner Kuiper Belt and severely reduced its population of cometesimals. Thus one side-effect of making the Oort Cloud is that Neptune probably cut short the formation of any other big planet farther from the Sun than itself. There just wasn't enough material left over to build it with.

It's a neat picture the computers have drawn. But is there any observational evidence for the Kuiper Belt? On August 30, 1992, David Jewitt and Jane Luu of the University of Hawaii found a dim object, since designated 1992 QB1. It orbits at a distance from the Sun of 44.3 AU in a period some 295 years long. (Recall that Neptune's average distance from the Sun is 30 AU and Pluto's is 40 AU.) 1992 QB1 proved to be the first of what are now over 200 objects with established orbits in the transneptunian realm. Calculations suggest that approximately 100,000 "QB1s" orbit in the region from 30 to 50 AU.

Assuming these objects have reflectivities like those of dark comet nuclei, their estimated diameters range from 60 to 600 miles, or 100 to 1,000 km. (Should they be more reflective, however, their sizes would be smaller.) About half of them appear grey in color, the rest somewhat reddish. A grey color hints at a surface that is either freshly covered in carbonaceous dust or is icy, but very old and heavily "weathered" by ultraviolet light. A reddish color suggests an icy surface rich in hydrocarbon compounds and moderately weathered. Yet exactly what the two colors are telling us about the histories of these distant objects is so far unknown.

Keep in mind that 1992 QB1 and all the others are simply the tip of the transneptunian iceberg. (The same goes for the handful of known Centaurs.) Below the limit of what telescopes can detect presumably lie thousands or millions more such objects, either too small to be detected as yet or simply far enough from the Sun to be lost against the blackness of space. The picture will doubtless grow more complex as searches push deeper into this remote realm.

Like all comets, Hale-Bopp formed from the same cloud of interstellar gas that the Sun and planets came from. Littering the Milky Way lie many similar clouds, like the North America Nebula in Cygnus (seen here). Glowing red with ionized hydrogen, such clouds are the galaxy's star-making "factories." (Tony and Daphne Hallas/Astro Photo; 6 x 7 cm format, 165 mm f/4 lens, 10 minutes on Kodak PPF 400 film; March 4, 1997)

For example, the known objects show that most Kuiper Belt bodies travel in orbits that are not too eccentric and have low inclinations to the ecliptic. But already there are hints that the region also contains a second population of objects with orbits that are much more eccentric and have higher inclinations. One recently discovered transneptunian body — 1996 TL66 — follows a highly eccentric, 800-year orbit that differs greatly from the other known Kuiper Belt inhabitants. Computer modeling shows that 1996 TL66 was probably pulled from the Kuiper Belt population by Neptune ages ago and then fired back into it, where it pursues its unusual path. Since 1996 TL66 is unlikely to be unique, scientists suspect a group of similarly scattered icy objects awaits discovery.

Interestingly, planetary scientists say that we caught our first glimpse of a Kuiper Belt object as far back as 1930, with the discovery of Pluto. This oddball planet and its moon Charon are unlike any other, except perhaps Neptune's equally odd moon, Triton. The three are built from similar mixtures of rock and ice, and Pluto and Triton have thin atmospheres that are cometary in the sense that they are slowly escaping into space. All three are large by comparison with 1992 QB1: Pluto's diameter is about 1,500 miles (2,400 km), Charon's is 750 miles (1,200 km), and Triton's is 1700 miles (2,700 km). This is about 10 times the average size of the transneptunian objects found to date. Yet, as scientists gain a better idea of the population and sizes of the Centaurs and the objects in the inner Kuiper Belt, these three bodies — Pluto, Charon, and Triton — seem to be fitting in well enough. They are big, but you might expect to have some giant icy planetesimals wandering around in the Kuiper Belt and at least a couple could survive until today.

Triton was captured by Neptune, but Pluto and its moon escaped that fate, as well as the dispersed fate of most Kuiper Belt objects. Pluto and its moon still circle the Sun because by happenstance the pair fell into an orbital lock-step with Neptune. This keeps them from ever approaching Neptune closely. About 35% of the known transneptunian objects have similarly protected orbits, while the remainder tend to stay well out of Neptune's reach. Triton, Pluto, and Charon are thus remnants, a small handful of big survivors left from the Kuiper Belt's original throng. One wonders how many other Plutos and Tritons still orbit in the Kuiper Belt and inhabit the Oort Cloud — and how many more took the course that the Centaurs will someday follow, slipping into orbits that lurch among the inner planets.

The inventory of the outer solar system has really only just begun, but already the objects found there are showing scientists that some of the old categories are too rigid. Witness Pluto — at once the last of the planets and the first of the Kuiper Belt objects. What scientists are looking forward to is filling in the gaps, both in space and in varieties of objects. It's an exciting time in the field, and if no one knows exactly where it's all leading, there's no question that comet science has come a long way from Aristotle's "dry exhalations."

Let's now see what it takes to turn one of these icy planetesimals into a Great Comet.

Comet Names and Discoveries

Once upon a time, comets didn't have names. They were called the "Great Comet of 1556," or whatever the year was when they were prominently seen. This was an era when comets were viewed as fearful omens, and understandably nobody wanted his or her name attached to the dreaded object. Moreover, since comets were portents sent by God as a warning to erring humanity, it would border on sacrilege to claim the comet, assuming anyone was bold enough to take such a step.

These superstitions began to fade as the Scientific Revolution took hold during the 1600s. By the time of Newton and Halley toward the end of the century, attitudes were clearly changing. As comets lost their aura of fear, at least among the educated, astronomers began systematically looking for them in hopes of testing theories. Another factor contributed to the shift in views. In an era when aristocratic patronage paid real dividends, astronomers found it worth their while to become known as comet finders. Honored positions and awards went to those who could add luster to a monarch's court, and scientific discovery was one of the newly approved ways for sovereigns to demonstrate to the world how advanced they were.

Thus when we look over the Great Comets of the early eighteenth century, we find for the first time that they begin to bear the names of astronomers. Some namesakes were observers — comet-hunters who spent every clear night glued to telescope eyepieces as they scanned the skies, just like their modern descendants do. Their rivalry was keen, both for financial awards such as gold medals and for the intellectual challenge. Among the most prominent eighteenth century comet-hunters were Charles Messier (1730–1817) and Pierre-François Méchain (1744–1807), two whose competitive comet searches were legendary.

Other astronomers discovered comets by calculating orbits and looking for past comets that might have been earlier visits from a single object. This was the route followed by Edmond Halley and Johann Encke. Today, orbit calculation has become a trivial task thanks to computers. But in a time that offered few mechanical aids to computation, working out an orbit cost many tedious hours of relentless arithmetic, and its practitioners unquestionably earned their heroic stripes.

By the middle of the nineteenth century, the invention of the telegraph meant that news of discoveries could be flashed to many observatories at once, and the pace of discovery picked up. Simultanously, astronomers evolved a system for designating comets that persisted almost unchanged down to 1995. Because this system is often encountered in older books, it's worth describing briefly before mentioning the system now in use.

In the old system, when a new comet was found it received a temporary designation, for example Comet 1982h. The designation shows that during the year 1982 this comet was the eighth (a was the first) to be discovered or recovered in case of already known periodic comets. Then once the comet had passed perihelion (perhaps within the discovery year or perhaps not), it received a permanent designation consisting of the perihelion year and a Roman numeral for its order in the perihelion sequence of comets for the year. In this particular case, Comet 1982h didn't come to perihelion until March 1986, and because it was the third comet to do so that year, its permanent designation thus became 1986 III. (But we know it much better as Comet Halley.)

Starting in 1995 the International Astronomical Union (IAU) began a new system because the old one was proving unwieldy. After-the-fact discoveries of long-gone comets on old photographic plates and archived satellite images broke into the established perihelion order for certain years, and improved observations and recalculations sometimes forced a shuffle in the lineup as well. Moreover, the IAU wanted a system that could handle a lot of discoveries in one year. The system also had to apply equally well to asteroid and comet discoveries, in part because at the time of discovery it's not always immediately clear which kind of object has been found.

So the rule now is that new discoveries (and recoveries) receive a designation made up of the year followed by a capital letter for the half-month in which it is found. A = January 1 to 15, B = January 16 to 31, C = February 1 to 15, and so on down to Y = December 16 to 31. (The letter I is omitted to spare everybody confusion.) This is followed by a numeral showing the order of announcement of discovery within that half-month. For example, Comet Hyakutake is designated C/1996 B2, because it was the second comet announced in the second half of January 1996. Similarly, Hale-Bopp is C/1995 O1 because it was the first comet discovered in latter half of July 1995. The "C/" indicates that it's a comet of long period (greater than 200 years). Had it been of shorter period, it would bear the designation "P/" for periodic.

Some periodic comets — Halley's, for instance — have been known for so long and their observational record is so good that they have very well determined orbits and have been observed at many returns. These comets graduate, so to speak, to a more renowned class in which they receive a sequential periodic-comet number. For example, Halley's Comet is designated 1 P/Halley, in honor of being the first comet to have its return successfully predicted. The second such comet is 2 P/Encke — not a Great Comet at all, but one which we'll be seeing more of thanks to upcoming spacecraft visits. In the 1996 edition of the definitive *Catalogue of Cometary Orbits*, the numbered periodic comets run as high as 124 and the tally has grown steadily with each new edition.

But what about the names of discoverers? Customarily, comets have borne the names of up to three independent discoverers, listed in order of actual observation. Thus we have comets such as Comet West, Comet Seki-Lines, or Comet Kobayashi-Berger-Milon. While nearly all comet discoverers are creatures of flesh-and-bone, not all are. For example in 1983, a small comet came close to Earth. It was picked up first by an orbiting observatory, the InfraRed Astronomy Satellite (IRAS), then by two amateur comet-hunters, Genichi Araki of Japan and George Alcock of England. The result was Comet IRAS-Araki-Alcock (and a fine sight it was, too!). More recently, the Solar and Heliospheric Observatory satellite (SOHO) discovered a slew of Comet SOHOs, all of them passing close to the Sun, where some were destroyed.

The arbiter of discovery claims is Brian Marsden of the International Astronomical Union's Central Bureau for Astronomical Telegrams. This is headquartered at the Harvard-Smithsonian Center for Astrophysics in Cambridge, Massachusetts. As one can imagine, sorting out discovery claims can involve touchy feelings at times, and there have been cases where none of the discoverers feels entirely satisfied. Marsden readily admits that the job puts him on the spot occasionally. But every sport needs a referee, and comet discovery is a centuries-old game whose participants certainly keep score.

Recently, Marsden proposed dropping the discoverers' names from comets altogether. This raised a howl of protest from many astronomers, and not just the amateurs, who had the most glory to lose in such a move. Many felt that if discovering comets is important at all — and no planetary scientist questions this — then it makes sense to keep alive a system of incentives that spurs people to hunt for them. Nor are the rewards completely intangible. In 1998, using a bequest by a friend of amateur astronomy named Edgar Wilson, the Smithsonian Astrophysical Observatory began administering a financial prize to be given to all amateur astronomers who discover comets each year. The Wilson Prize award, $20,000 for the first year, is to be shared equally by all amateur comet discoverers on a per-person, per-comet basis.

Whether or not money is at stake, the competition to discover comets can be as fierce as anything back in the eighteenth century. To pick just one example, as Comet Halley was approaching during the early 1980s, several teams of astronomers were vying to be the first to detect it since it was last photographed on June 16, 1911, over 70 years earlier. On October 16, 1982, Ed Danielson and Dave Jewitt won the race using the 200-inch telescope at Palomar. They spotted Halley at 24th magnitude at a distance of 11 astronomical units from the Sun.

And what about the also-rans? The team that came in second to recover Halley missed beating Danielson and Jewitt by only *3 hours*.

That's competition.

2 *What Makes a Comet "Great"?*

Many big and bright comets have acquired the nickname "Great," but until recently no astronomer bothered with a formal definition. In a sense, definitions don't matter — any comet that impresses the onlooker counts as a Great Comet. But still it's interesting and perhaps scientifically useful to take previous Great Comets and see if they possess anything in common besides the name. That is what David Hughes of Britain's University of Sheffield did about ten years ago. He analyzed observations of historic Great Comets and distilled a set of qualities they all displayed. A similar set of criteria emerged from studies by Donald Yeomans of the Jet Propulsion Laboratory in California.

In particular, Hughes pinpointed five criteria that have to be met for a comet to rank as a Great Comet. Such a comet must:

- have a large nucleus and coma,
- have a large active surface area,
- reach perihelion near the Sun,
- pass close to Earth,
- provide good viewing opportunities for terrestrial observers.

To these Yeomans added how long the comet is visible to the unaided eye in a dark sky. He used this criterion to filter out comets that become bright but are always seen near the Sun, which greatly reduces their spectacle.

Luck obviously plays the controlling role in choosing which comets will be great ones. As we'll explore in more detail below, a comet can pass very near Earth, as Comet IRAS-Araki-Alcock did in May 1983, yet not reach Great Comet status because it is too small or too inactive. And a large comet may never get very close to the Sun and yet still be a Great Comet, as the Great Comet of 1811 did, thanks to its large size and high rate of activity. Moreover, even normally bright comets that come close to the Sun and Earth can be placed at a disadvantage during a particular pass through the inner solar system due to accidents of timing and

geometry — as happened with Comet Halley in 1986. Each and every comet can almost amount to a case unto itself.

Let's now take a look at each of the criteria in more detail.

Large nucleus and coma

As we saw in Chapter 1, comet nuclei range greatly in size, from perhaps a few dozen yards or meters across to several hundred miles if we include such bodies as 2060 Chiron. But unfortunately, the database of accurately measured comet nuclei is woefully small. Photos showed that the nucleus of Halley's Comet stretches 10 miles (16 km) in its longest dimension, and the best estimate for Comet Hale-Bopp yields a diameter of 25 miles (40 km). Beyond that, however, we have little more than educated guesses: Comet Wirtanen seems to have a nucleus 0.8 mile (1.2 km) across, the heart of Comet IRAS-Araki-Alcock is 4 to 8 miles (6 to 12 km) in diameter, and Comet Bennett may have a nucleus 5 miles (8 km) across.

A large nucleus contributes to making a comet great because the comet's brightness depends critically on how much gas and dust it releases, and a large nucleus has a greater supply of both. The gas and dust from a large comet will create a larger coma, which shines both by its gases fluorescing and by reflecting sunlight from the dust. The bigger the coma the brighter the comet, all else being equal. A typical coma can reach a diameter of several hundred thousand miles, but significantly larger ones have occurred. For example, the Great Comet of 1811 had a diameter measured at over a million miles' across — or some 50% larger than the Sun itself.

As a comet nucleus sails into the inner solar system, the coma it produces grows with it, usually reaching its maximum size around the orbit of Mars. This is roughly where the increasing power of sunlight overpowers the growing internal pressure of the coma. In effect, this is where a large, "soft" coma begins to transform into a smaller, "harder" one. Yet, despite shrinking its size, the coma's brightness increases. This occurs because the increasing warmth means more outgassing and thus more fluorescence. Likewise, the dust is thicker hence more reflective, and the sunlight it is reflecting is stronger too.

A flashlight lit the rocks as Comet Hale-Bopp stood overhead at Joshua Tree National Park in California. (Wally Pacholka; 50 mm f/2 lens, 30 second exposure on Fuji 800 film; April 4, 1997)

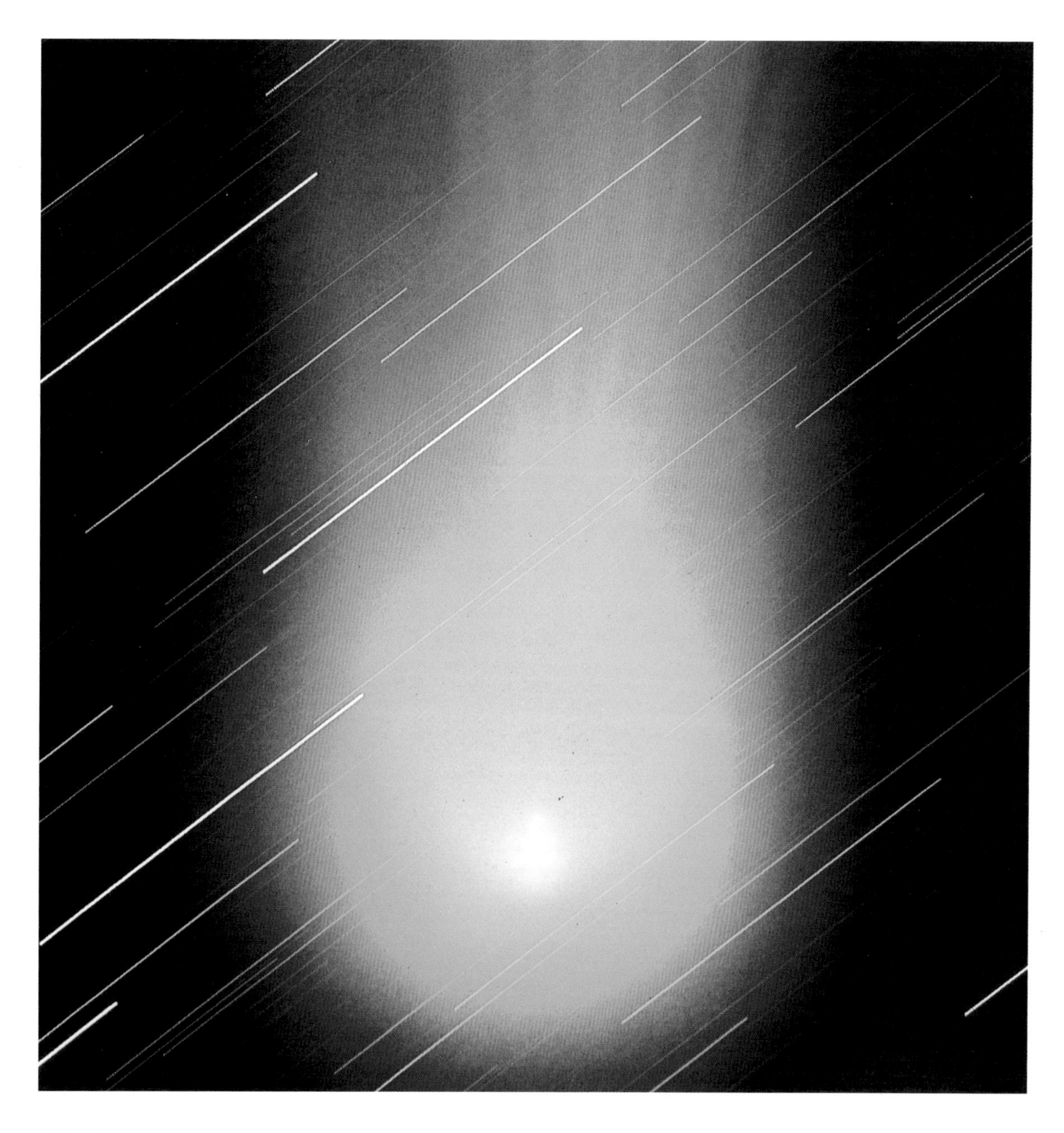

Although Comet Hyakutake wasn't large, it came close to Earth. This turned a comet that might have been ordinary into a Great Comet that was outshone only by Hale-Bopp a year later. (Tony and Daphne Hallas/Astro Photo; 7-inch f/7 Astro-Physics refractor, 30 minutes on Fuji SuperHG 400, using SBIG ST 4 autoguider; date not recorded)

In the summer of 1957 Comet Mrkos appeared, the second bright comet of that year, following Comet Arend-Roland. Here its gas tail betrays turbulence in the solar wind (Palomar Observatory/Caltech; 7 minute exposure with 48-inch Schmidt camera; August 22, 1957)

An Active Nucleus

Just as a large nucleus provides much raw material for a Great Comet, a nucleus that is active over much of its surface will likely be in the running for Great Comet status. Studies of many comets show that most have remarkably little surface area active — the average for most comets lies under 10% and this can even drop to 1% of the surface for old, well-traveled comets. Comet Halley, in contrast, appears to be unusually active, with 10% to 20% of its surface emitting gas. By comparison, Comet IRAS-Araki-Alcock has a nucleus roughly as big as Halley's, but much less active. This implies that most of its surface is thickly mantled by dust and easily heated patches of ice and snow are few. (Alternatively, it may have evaporated most of its frozen material.)

In Halley's case, and with many other comets, much activity takes the form of individual jets. These are places perhaps a few hundred yards or meters across where the comet's ice is laid bare. Exposed to sunlight and warmth, the icy material produces a geyser of vapor that at full strength can shoot out at roughly 1 kilometer per second, the muzzle velocity of a high-power rifle bullet. Great quantities of dust are also caught up in these outflows because the dust permeates the ice and comes loose as the ice evaporates. Since sunlight falling on the ice is the only source of heat, jet activity typically turns on slowly at local sunrise, grows to maximum strength in the early afternoon (when solar heating reaches its peak), and dwindles to inactivity around local sunset. (Hollywood exaggerated this in *Deep Impact*: jets don't erupt as abruptly as the movie showed or with such landmine force.) The pronounced daily cycle, however, gives a useful way to determine the rotation period of the unseen nucleus — simply time the recurring changes in the comet's innermost coma.

Hale-Bopp was exceptionally active, which contributed strongly toward making it a Great Comet. Here it is passing near the Dumbbell Nebula (M27) in Vulpecula. (Alessandro Dimai, Orsola and Carlo Ferrigno; Takahashi 102 mm f/6 telescope, 8 minutes on hypered Kodak Express Gold 400 II; February 10, 1997)

Jets can also shuffle the dust on the surface of the nucleus. As the nucleus rotates, the heavier particles that have been lifted by the jet but not accelerated to escape velocity will fall back to the surface and accumulate on the "western" side of the jet's active area. (For the sake of illustration let's assume that all comet nuclei rotate toward the east.) While the gravity of a comet nucleus is extremely feeble – you could easily jump off a comet, since the velocity of escape is only about 1 mile per

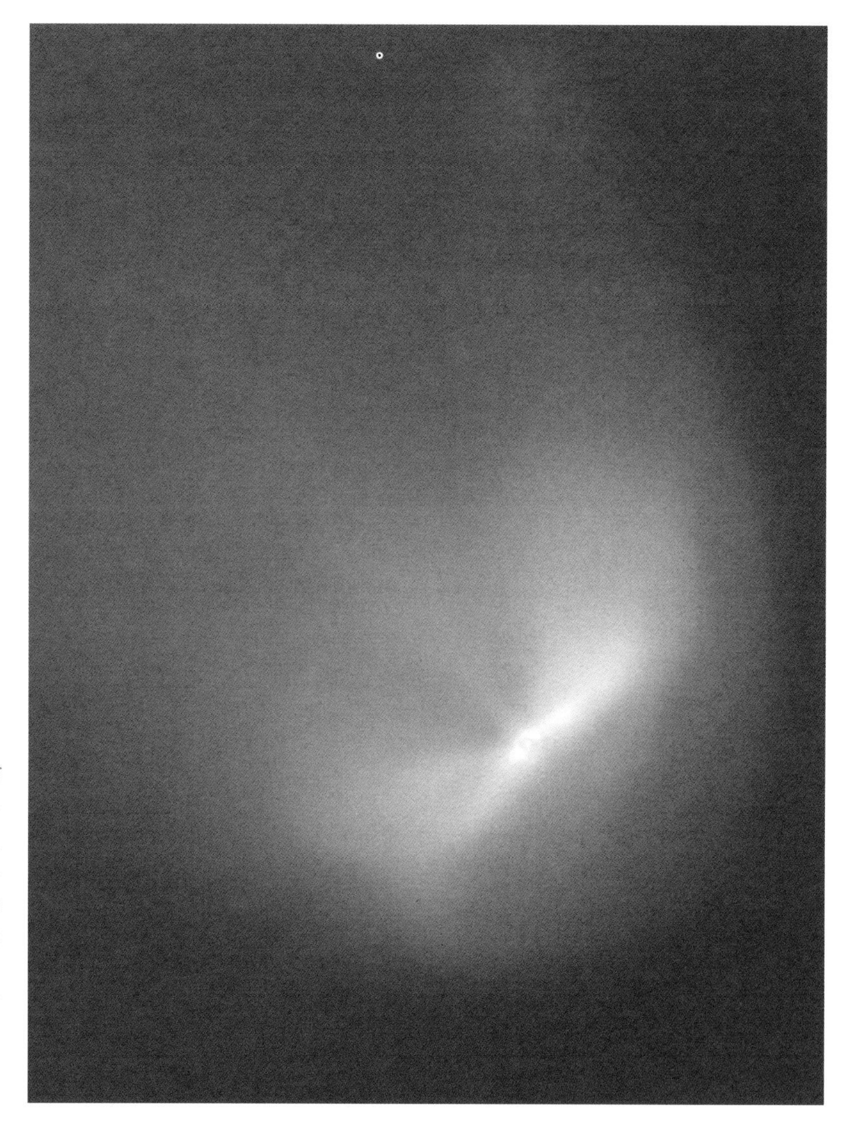

An outburst on the nucleus of Comet Hyakutake was photographed on April 15, 1996, by a team at the Pic du Midi Observatory, France. A similar outburst on Hale-Bopp shortly before it was found may have brightened the comet and made it easier for Alan Hale and Tom Bopp to discover it. (Observers: F. Colas, J. Lecacheux, P. Laques)

hour — eventually a growing pile of dusty debris will accumulate to one side of the jet. At some point, this will become unstable and topple. Some dust will likely avalanche back across the fresh surface of the active jet and smother it, at least temporarily. Observers with ground-based telescopes have witnessed changes in jet activity in most comets, and this mechanism could be the cause. Such avalanches — not to

Although Hale-Bopp did not pass very close to the Sun, its large and active nucleus easily transformed it into a Great Comet. (Steve Padilla and Ron Royer; April 9, 1997)

Sungrazer — Great Comet Ikeya-Seki rose tail-first out of the dawn in October 1965. The comet belongs to a family of sungrazers that resulted when a giant comet broke up in prehistoric times. Ikeya-Seki passed very close to the Sun and its nucleus split into three pieces (invisible here). One piece quickly evaporated, but the other two are heading outward on slightly different orbits. The yellowish color comes from strong sodium emission. (Roger Lynds, National Optical Astronomy Observatories; 4 minutes on Kodak High-Speed Ektachrome; October 29, 1965)

mention jets, a steady rain of dusty hail, and blizzards of icy cobbles and small fluffy boulders — pose concerns for spacecraft designers planning to put landers on comet nuclei.

As pockets of exposed icy material become exhausted, activity there will slow or stop, while new areas may begin erupting as solar heat creeps through the dusty mantle and triggers the evaporation of fresh ice. Since a comet nucleus is rough and irregular at all scales, tectonic movements such as landslides or the collapse of elevated hills may abruptly expose new areas of ice or fresh deposits of dust. This holds especially true after perihelion, when the nucleus has been hit by a strong dose of solar heat and the surface materials are probably the most disturbed. Many comets are notably brighter and dustier after perihelion than before.

While jets are the most spectacular form of nucleus activity, significant amounts of gas may leak through the dusty mantle wherever it has not accumulated too thickly. This activity, plus the evaporation and breakdown of dust particles into their constituent atoms and molecules, produces a general foggy haze that merges imperceptibly with the rest of the coma and ends up in the ion tail.

Perihelion near the Sun

A Great Comet is above all an active comet, and this means getting near enough to the Sun that its heat can kick the comet's activity into high gear. While a thin coma starts to develop when a comet is far out beyond Jupiter, the real action usually commences once it comes within about 3 astronomical units (AU), where water ice begins to turn to gas. In practice, all Great Comets should reach perihelion at a distance of less than 1 AU, the average distance of Earth's orbit – but many of the best Great Comets have come much closer to the Sun than that. Here are some figures:

Comet	**Perihelion distance**
Comet Bennett (1970)	0.54 AU
The Great Comet of 1618	0.40 AU
Comet West (1976)	0.20 AU

The Great Comet of 1577	0.18 AU
The Great January Comet (1910)	0.13 AU
The Great Comet of 1680	0.01 AU

Some have gone even closer: the Great Comet of 1882 got down to 0.008 AU, and the Great March Comet of 1843 reached 0.006 AU. This last comet skimmed only 81,000 miles (131,000 km) above the Sun's surface – the Sun's radius is 0.00465 AU – and produced a tail 190 million miles (300 million km) long, far enough to stretch beyond the orbit of Mars.

One bright comet stands as an exception to the "go close" rule — the Great Comet of 1811, whose perihelion lay at 1.04 AU. But this comet was intrinsically large and extremely active, with the result that it was outstanding even though it never came very near to the Sun. The Great Comet of 1811 remained visible to the naked eye in a fully dark sky for nearly 9 months — longer than any comet before or since except for Comet Hale-Bopp, which was visible to the unaided eye for an astounding 15 months.

Comets that sail very close to the Sun are called sungrazers and they form a select group. How close can a comet get to the Sun? Some have actually struck it and perished. Astronomers have occasionally seen comets race in toward perihelion and fail to emerge, and from time to time a comet has been glimpsed near the Sun during a total solar eclipse but never seen again. These comets must have either hit the Sun or been evaporated by its heat. Such sightings, however, are entirely fortuitous and have always been scanty in number: up to 1979 less than a dozen comets had ever been recorded close to the Sun.

In the past 20 years, however, spacecraft observations have made the records far more complete. Orbiting solar cameras have spotted many comets passing near the Sun. Using an instrument named Solwind, an Air Force test satellite in Earth orbit discovered six sungrazing comets between 1979 and 1984, and in 1979 and 1981 it photographed three of them taking fiery plunges into the Sun. These comets were found by happenstance when Solwind was imaging the Sun's corona. By 1989, a different solar monitoring satellite named Solar Max had caught 10 additional comets that either struck the Sun or went too close to survive.

More recently, the Solar and Heliospheric Observatory (SOHO) spacecraft, launched in 1995, watched as more than 60 comets took a fast (and usually fatal) passage close by the Sun.

With only a few exceptions, the comets caught by spacecraft have been small, having nuclei probably no more than a few hundred yards across. They are bright only because they lie within a couple of solar radii of the Sun's surface. If one of them were to pass by Earth at the distance of Comet Hale-Bopp, for instance, we might never even discover it. Perhaps the most remarkable thing about these comets is that almost every one of them was detected *solely* by the satellites' automated cameras. This fact is somewhat worrying because it suggests that current comet searches, which naturally focus on the nighttime sky, are overlooking a lot of comets that approach perihelion on paths that keep them hidden in the glare of light from the Sun. There are clearly many comets we're not catching, and conceivably one of them could come at us straight out of the Sun. We'll return to the subject of risks from comets in Chapter 7.

The crowd of sungrazers contains many small objects. Yet it also includes a few Great Comets, most notably Comet Ikeya-Seki of 1965. Ikeya-Seki passed within 0.008 AU of the Sun at perihelion, or less than two solar radii above the surface. With sunlight beating on the comet many thousands of times more strongly than at Earth's distance, the comet grew so hot that astronomers detected emission from atoms of vaporized iron in the comet's dust. Other metallic atoms found in Ikeya-Seki included cobalt, potassium, nickel, and manganese. (Presumably, had modern instruments observed the even-closer passage of the Great March Comet of 1843, similar results would have been found.) Ikeya-Seki didn't escape unharmed from the close passage. Immediately after perihelion its nucleus broke into three pieces, one of which quickly vanished as it evaporated. The two remaining nuclei departed on slightly different orbits and will return in periods of 880 years and 1,056 years respectively.

The breakup of Ikeya-Seki surprised few astronomers, as the comet is a member of a group of related sungrazing comets called the Kreutz family, after Heinrich Kreutz (1854–1907) who identified the first few members by their similar orbits. More recently, work by other scientists

One Great Comet broke up even without going extremely close to the Sun. Comet West separated into at least four fragments after perihelion, when activity on its nucleus was strongest.

THIS PAGE
Steve Knudson, Malcolm B. Niedner, Joint Observatory for Cometary Research — NASA Goddard Space Flight Center and New Mexico Insititute of Mining and Technology [*whole comet*, April 1, 1976].

FACING PAGE
Elizabeth Roemer, Catalina Station, Lunar and Planetary Laboratory, University of Arizona [*nucleus closeup*, April 25, 1976].

(mainly Brian Marsden) has reconstructed a history for this remarkable family of comets. The original parent comet may have had a nucleus about 75 miles (120 km) in diameter, and it seems to have arrived in the inner solar system between 10,000 and 20,000 years ago. Whether this was its first trip from the Oort Cloud or the result of perturbations by Jupiter or some other planet is unknown. In any case, the mega-comet wound up in an orbit that skimmed the Sun every 1,000 years or so. As

Marsden figures it, roughly 10,000 years ago the Kreutz parent comet broke apart on one of its sungrazing passages. The rupture spawned two major fragments traveling on similar but not identical orbits.

The first Kreutz fragment held together until 371 BC, when it broke into at least three large pieces, one of which became the Great Comet of 1843. The second Kreutz fragment remained intact until AD 1106, when it split into three or more comets. Its decendants include the Great

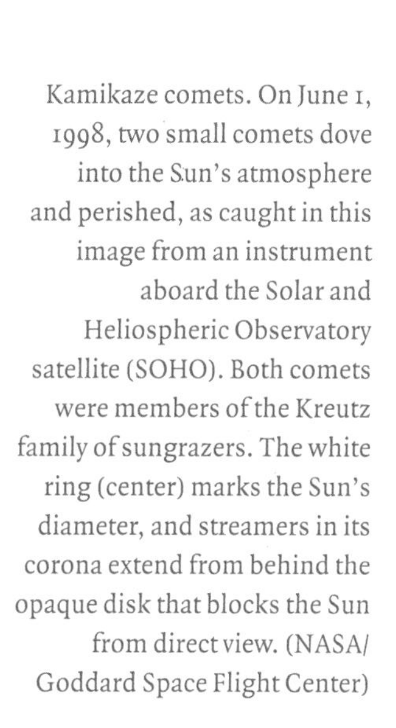

Kamikaze comets. On June 1, 1998, two small comets dove into the Sun's atmosphere and perished, as caught in this image from an instrument aboard the Solar and Heliospheric Observatory satellite (SOHO). Both comets were members of the Kreutz family of sungrazers. The white ring (center) marks the Sun's diameter, and streamers in its corona extend from behind the opaque disk that blocks the Sun from direct view. (NASA/ Goddard Space Flight Center)

Comet of 1882 and Comet Ikeya-Seki of 1965. Many additional Kreutz member-comets are now known or suspected, including the first Solwind comet and more than 50 of the sungrazers found by SOHO. These all result from the same process, because breaking up a comet appears to be rather like smashing a vase. In both cases, you're left with a number of big pieces, lots of smaller fragments, and uncountable myriads of tiny ones.

While not a Kreutz-family sungrazer, another disrupted Great Comet was Comet West of 1976. It passed within 0.2 AU of the Sun and broke into at least four pieces. While they attribute the breakup of Ikeya-Seki and the Great Comet of 1882 to solar tidal forces, scientists are more inclined to finger massive gas eruptions as the predominant force in shattering most comet nuclei. And if West was not subjected to heating

All the ingredients have to be there... Comet IRAS-Araki-Alcock passed quite close to Earth in May 1983, but because the comet's nucleus was relatively inactive, it never became a Great Comet. (Marc W. Buie; 205 mm f/3.8 lens, 30 minutes on Kodak VR 1000 film; May 9, 1983)

The Un-Great Comet of all time — Comet Kohoutek — produced a good show for observers who saw it under clear skies. Yet, for reasons poorly understood at the time, it didn't come through as the "comet of the century." (Robert G. Roosen and John C. Brandt, Joint Observatory for Cometary Research – NASA Goddard Space Flight Center and New Mexico Insititute of Mining and Technology; January 13, 1974)

as strong as that experienced by Ikeya-Seki, at perihelion it nonetheless felt a blast of sunlight 25 times stronger than at Earth's orbit. If the nucleus had any inherent weaknesses or fault lines (which is quite plausible), powerful eruptions of gas could have done the trick.

Passes close to Earth

If an ordinary comet happens to pass near Earth it can become a Great Comet simply because it will appear bright and large. But a close passage is not enough in itself. For example, the prize for the closest known cometary approach to Earth goes to Comet Lexell of 1770, which missed us by only 1.4 million miles (2.2 million km), less than 6 times the distance to the Moon. Despite coming so close, Lexell's Comet never developed much of a tail and its coma spanned no more than about $2^1/_2°$ across — five times the apparent size of the Moon. Another close passage by a non-Great Comet was IRAS-Araki-Alcock. In May 1983, it swept past at a distance of 12 times that of the Moon, or 2.7 million miles (4.6 million km). Although it never grew a long tail or became very bright, Comet IRAS-Araki-Alcock was exciting to watch. At closest approach during the second week of May, observers with telescopes could see the comet moving visibly across the stars like a piece of Milky Way set adrift.

One of the most recent Great Comets, Hyakutake, certainly benefited from a close approach to Earth. It flew past at a distance of 0.1 AU (9 million miles; 15 million km). The close distance turned a relatively small comet into a show-stopper. And conversely, while Great Comet Hale-Bopp passed 1.3 AU away from us, its large nucleus and great activity made up for what it lacked in proximity. (If Hale-Bopp had shot past at Hyakutake's distance, it would have been 150 to 200 times brighter.)

The formula for a Great Comet doesn't necessarily include a long tail — but it never hurts to have one. There is just something about a long glowing banner reaching across the sky that says "Great Comet" to many people. For example, the Great Comet of 1811 displayed a tail that stretched nearly 70° at maximum. Donati's Comet of 1858 was a particularly beautiful object with a curving dust tail that reminded some viewers of a scimitar as it stretched 60° across the sky. The Great Comet of 1861 had a tail reported to be longer than 90° at its closest approach to Earth, 14 million miles (23 million km). In 1910, the Great January Comet upstaged the appearance of Comet Halley a little later that same year. The January comet was even bright enough to be seen in broad daylight within a few degrees of the Sun. It flew past Earth at 0.9 AU, and showed a tail stretching 30° to 50°. And Comet West in 1976 came within 0.8 AU of Earth (75 million miles; 120 million km), developing a beautiful gas tail about 30° long, with many streaks in the dust tail.

A long tail could contribute strongly to the impression a comet makes on a fearful public. Comet writings in the Middle Ages and after frequently noted a comet's "frightful" appearance, and it seems clear that many people saw a long comet tail as a sword blade held to mankind's collective throat. Even today, when people in the developed world have a fair notion of a comet's real nature, it's hard to avoid a slightly spooky tingle when you first see one of these objects in the sky. So chalk up a long tail to the ingredients that helps to make a Great Comet.

Good viewing opportunities

A Great Comet has to be placed where people can see it easily, even if all the other criteria are met. David Hughes says that a comet should be at least 20° away from the Sun, and Don Yeomans addresses the same point by measuring only the time a comet is visible in a completely dark sky. Another factor helping a comet's fame is an orbit that favors the more populated Northern Hemisphere. Lastly, for understandable reasons, people find it much more convenient to go out and look at a comet in the evening sky than in the predawn, so those visible after sunset tend to attract larger audiences and greater reputations.

To pick an example, the 1986 return of Comet Halley broke several good-viewing guidelines and the result disappointed many onlookers. There's no question that this particular comet has the potential for being a Great Comet — after all, it has been a Great Comet on many of its returns. Halley contains plenty of material to produce lots of gas and dust. It is abundantly active, it reaches perihelion at 0.6 AU, and it can pass close to Earth. However, when Halley's Comet raced through peri-

When all the circumstances come out right, the result is a Great Comet remembered for years after. (Comet Hyakutake; Chuck Vaughn, 350 mm f/2.8 lens, two 10-minute exposures on Fuji SuperG 400 Plus film sandwiched together; March 24, 1996)

helion in February 1986, it lay on the other side of the Sun from Earth and was lost in the glare. Then as Halley rose to its post-perihelion peak of activity in March and zoomed closest to Earth in April, it was moving southward on the sky — great for observers below the equator, but giving poor views for the Northern Hemisphere, especially for those looking at it from cities where lights brightened the sky. (Moonlight also interfered during Halley's brightest period.) In the end, a truly wonderful comet — "mankind's comet," in the words of Guy Ottewell and Fred Schaaf — regrettably sank to just a mediocre apparition by comparison with others of recent memory.

Halley was a lot better in 1910 for our grandparents and great-grandparents, even considering the competition from the Great January

Comet of that same year. Halley was at its best in April and early May 1910, where it joined the planet Venus to make a celestial landmark in the predawn sky. Reaching perihelion on April 20, Halley lay well to the west of the Sun and was prominently visible before sunrise. On May 19, the nucleus passed directly across the Sun's disk (and showed that it was too small to be seen in silhouette). The next day the comet lay only 0.15 AU from Earth (14 million miles; 23 million km), having passed just inside our orbit. Then after shifting into the evening sky, the now-outbound Halley was positioned so people could spot it easily without the chore of rising before dawn, and this is when most people actually did see it.

Visibility can also be aided by the inclination of the comet's orbit to that of Earth. Comets like Hale-Bopp which have orbits that incline steeply tend to place the comet well away from the Sun during most of its apparition, naturally improving visibility. Hyakutake, whose orbit has a lower inclination, nonetheless favored the Northern Hemisphere as it cruised by Polaris, the North Star, when near peak brightness. This placed it perfectly for viewers across North America, Europe, and northern Asia, some of the world's most populated regions. And again, it was highly visible in the evening sky — indeed it was visible all night long.

Un-Great Comets

While the criteria noted by David Hughes and Donald Yeomans are common to Great Comets, they aren't totally foolproof since nature can sometimes throw a curve ball. Every now and then a comet arrives that fits the categories well enough but still doesn't live up to expectations. One famous Not-Great Comet of a generation ago still leaves planetary scientists smarting: Comet Kohoutek of 1973–74. Here's what happened. Based on the comet's unusual brightness while it was far from the Sun, scientists made glowing predictions for the display it would put on when it swept through the inner solar system months hence. However, the predictions fell dismally short of reality due to reasons that were unforeseeable at the time.

When discovered near the distance of Jupiter's orbit, Comet Kohoutek was making its first approach to the Sun in ages and ages, perhaps its first time ever. In effect, it was a fresh arrival from the Oort Cloud. What scientists did not anticipate back then was that its nucleus was coated with a layer of highly volatile frost. This material burst brightly into fluorescence while the comet was still a long way out from the Sun. Not understanding that they were watching the disappearance of a thin coating of frost, astronomers took the brightness at face value and interpreted it to mean that Kohoutek had a huge nucleus and coma. So they extrapolated to make a prediction that the comet would be a spellbinding sight — brighter than Venus — when it finally came near Earth. Unfortunately, they were caught short when the nucleus evaporated the frost and reverted to a more normal level of activity. In essence, Kohoutek had done everything a Great Comet should, but it "flopped" because its nucleus did not behave in the expected way.

Yet Comet Kohoutek's bad reputation is undeserved. Although the general public was greatly disappointed in the comet's visual appearance, the scientific results from its visit were outstanding, including the first-ever cometary detections of methyl cyanide (CH_3CN) and hydrogen cyanide (HCN) by radio astronomers, as well as the identification of silicon (quartz) grains in the spectrum of the comet's dust tail. Comet Kohoutek also taught scientists a useful lesson about handling the media and public expectations. This experience led them to keep predictions on the conservative side (perhaps too conservative, in retrospect) when Great Comet West appeared in 1976. But the caution stood scientists in very good stead with Comet Austin of 1990, which behaved precisely like Comet Kohoutek. Nor were the lessons forgotten with the more recent arrival of Great Comet Hale-Bopp, which initially shaped up to be yet another repeat of the Kohoutek saga. Thanks to the knowledge hard-won with Kohoutek, however, scientists knew a lot more about how comets behave when far from the Sun. And luckily, during the nearly 20 months' warning between the time of discovery and its best appearance, Hale-Bopp splendidly proved the skeptics wrong.

Let's now turn our attention to the two magnificant Great Comet visitors of the past few years.

When is a comet not a comet?

Few people, astronomers included, have much trouble identifying a comet when they see it in the sky. The long luminous tail, the bright glowing head — these are pretty distinctive hallmarks. Even a comet that's far from the Sun and displaying only a little activity is not hard to identify when seen in a telescope.

But upon closer scrutiny, this clear picture grows a little more murky. For example, some objects follow comet-like orbits but show no evidence of cometary activity. And other objects, traveling in asteroid-like orbits, have developed gas and dust comas and even display small tails. Finally, there is at least one inert object whose orbit matches that of a meteor stream that most probably is a ribbon of cometary debris shed by it ages ago. So what's going on?

As instruments and telescopes improve and spacecraft pay more visits to solar system targets, planetary scientists have found that some old categories don't fit the current knowledge without some tugging and pulling.

Take the case of Chiron. In 1977 astronomer Charles Kowal discovered an object whose orbit came inside that of Saturn and reached outward of Uranus. Although its orbit lay well outside the main asteroid belt, which is between Mars and Jupiter, the new object looked properly asteroidal — that is, it showed simply as a star-like point of light in a telescope. Assuming that it was about as reflective as other asteroids, planetary scientists estimated the body was 90 to 150 miles (150 to 250 km) across. It was named Chiron and given the minor planet designation number 2060.

However, study of its motion using computer simulations showed that Chiron's 51-year-long orbit is highly unstable and in fact is evolving inward from the Uranus/Neptune region. But that's not the weirdest thing about this asteroid. As astronomers followed Chiron's slow approach toward its 1996 perihelion, they noticed it was brightening faster than its decreasing distance from the Sun could account for. In 1988 Chiron began to sport a coma and eventually a short tail.

Surprise, surprise — minor planet Chiron is actually a comet.

In the generation since Chiron was discovered, nine more objects have turned up with unstable orbits like Chiron's, and all probably form a related class of object. Taking a tip from Greek mythology, scientists call the entire group Centaurs. (In antiquity these were imaginary creatures with the torso and head of a man attached to the body and legs of a horse; Chiron was reputedly the wisest and best of them.) Although the other Centaurs don't appear to share Chiron's cometary activity, scientists think the group members are all icy planetesimals (or big comet nuclei, in other words) that have recently moved in from the Kuiper Belt beyond Neptune and Pluto (see Chapter 1).

But Chiron isn't alone in being an active asteroid. In 1992, scientists realized than an ordinary-looking asteroid discovered in 1979 was the same object as one discovered in 1949. But back in 1949, this object, which circles the Sun every 4.3 years, had displayed a small coma and tail. That proclaimed it to be a comet, so it was duly named Wilson-Harrington after its two discoverers. Following its rediscovery as an asteroid, Comet Wilson-Harrington is now also known as minor planet 4015 Wilson-Harrington. (This object may be visited by the Deep Space 1 probe early in the next decade — see Chapter 5.) Scientists are eager to get a look at Wilson-Harrington because one theory for why it appeared cometary in 1949 says that the coma seen at the time of discovery was a cloud of dust surrounding the body in the aftermath of a then-recent meteorite impact. If true, then Wilson-Harrington would indeed be just an ordinary asteroid.

Other odd objects include minor planet 5335 Damocles, discovered in 1991. It has an elongated, high-inclination orbit that appears like that of a short-period comet — except Damocles stubbornly refuses to get active and seems completely inert. Another case of the same behavior is 1996 PW. This object reaches perihelion in the middle of the asteroid belt and has a 6,000-year-long orbit that resembles that of a long-period comet from the Oort Cloud. Yet it shows a bare surface and no cometary activity. Scientists surmise that 1996 PW is an old comet nucleus whose icy materials are completely smothered by dust.

Something similar appears to have happened to the minor planet 3200 Phaethon. This asteroid was detected in data taken by IRAS, the InfraRed Astronomy Satellite, in 1983. When planetary scientists located Phaethon with ground-based telescopes and computed an orbit for it, they were surprised to discover that it travels closer to the Sun — 0.14 AU — than any other known minor planet. Even more startling, its orbit matched the orbit of the Geminid meteor shower. Since meteor showers are believed on good grounds to be cometary debris, this strongly suggests that Phaethon is really a extinct comet nucleus.

Centaurs, comet-asteroids, and asteroid-comets — each type occupies part of a spectrum that scientists are only beginning to explore. And understanding where these oddballs fit into the emerging picture is helping us get to grips with the planetary system we call home.

3 Great Comet Hyakutake (1996)

Even when an astronomer sets out to find a comet, the actual discovery always comes as a surprise. However, few comets give their lucky discoverers the kind of jolt that Comet Hyakutake (1996 B2) did in January 1996. The amazing story actually began the month before with the discovery of a totally different Comet Hyakutake (1995 Y1).

As dawn was about to begin on December 26, 1995, Japanese amateur comet-hunter Yuji Hyakutake was scanning star fields. He was looking for comets just as he had been doing on countless nights since he began to search in earnest the previous July. The hunt this particular morning was shaping up to be as fruitless as all the others had been. But Hyakutake, a commercial lithographer in his mid-40s, wasn't too discouraged. He was using his sole piece of observing equipment, a pair of 25 × 150 Fujinon binoculars. These resemble ordinary binoculars but they are gigantic, with main lenses 6 inches in diameter. They are essentially two low-power telescopes yoked together. And if they had yet to catch a comet for Hyakutake, they still provided beautiful views of the universe as he swept them methodically over the sky, looking for something that wasn't part of the normal backdrop of stars, nebulae, and galaxies.

On this morning, however, the routine was about to be broken. At 5:40 a.m., shortly before he planned to call it quits for the night, Hyakutake spotted a small fuzzy object where the constellations of Libra, Hydra, and Virgo meet. The hazy object was about 4 arc-minutes across, shining dimly at magnitude 10.5. To Hyakutake, it certainly looked like a comet, even though the fuzzball showed no hint of a tail and its northeastward motion was too slow for him to verify in the time remaining before dawn. But no galaxies or star clusters of that size and brightness lurked nearby, so the likelihood of a false alarm was small. Still, Hyakutake played it cautious — after all, this was his first comet discovery and he didn't want to make a fool of himself. So he hedged

While passing closest to Earth, Comet Hyakutake was moving swiftly enough that getting sharp photos required tracking the comet carefully. This drew out stars into streaks. (David Churchill; 4-inch Televue Genesis, f/5.4, 30 minute exposure on hypered Fuji 800; March 23, 1996)

his bets and reported it only as a probable comet to Japan's National Astronomical Observatory.

To his delight, staff astronomers soon confirmed the discovery and passed his positions for it along to the International Astronomical Union's Central Bureau for Astronomical Telegrams, the clearinghouse for all kinds of astronomical discoveries. It is located at the Smithsonian Astrophysical Observatory in Cambridge, Massachusetts. Using Hyakutake's observations plus those of others, the Central Bureau computed an orbit for the new comet and sent out a worldwide notice by e-mail, designating it as comet C/1995 Y1 (Hyakutake). Yet because of the date when he found the comet, Hyakutake thought of it as his "Christmas comet."

As January 1996 began, Hyakutake's Christmas comet continued its leisurely course toward the northeast, growing slightly brighter. By late February 1996, when it lay closest to both the Sun and Earth, it reached a maximum brightness of about magnitude 8.7, too faint to be seen by the naked eye, but easy in small telescopes. After the end of February, the comet slowly faded as its distance increased.

But this faint Comet Hyakutake was not "the" Comet Hyakutake. The Great Comet of 1996 was Hyakutake's second discovery, and it was found just five weeks after the first, on January 30, 1996, once again at around 5 a.m. local time.

As before, Hyakutake was out at his favorite dark-sky observing site, about 10 miles from the village of Hayato, where he lives in the southern Japanese prefecture of Kagoshima. On this occasion, following a lengthy run of bad weather, he hoped to photograph his Christmas comet. He knew where it was that morning, but clouds were scudding across that part of the sky. Through gaps he could glimpse it, but the opportunities for photography were nil. Nevertheless he decided to make the most of the sky conditions and continued comet-hunting with his massive binoculars. As it happened, the shifting breaks in the clouds led him back to the same spot among the constellations — between Libra and the tail of Hydra — where he had originally found his comet on the day after Christmas.

But when he looked into the binoculars this time, Hyakutake was startled and confused. The binocular eyepieces showed him a dim,

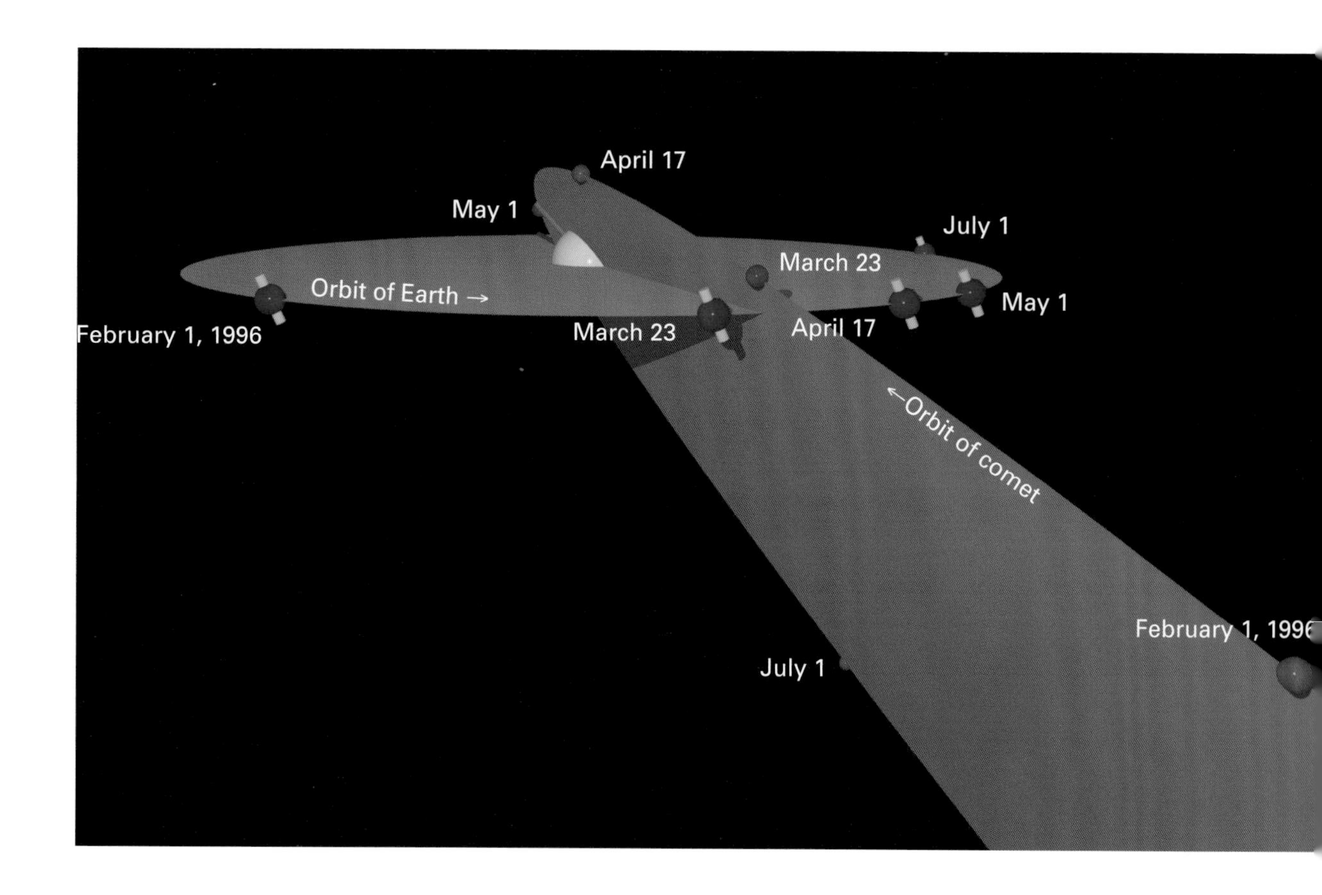

Flying visit. Comet Hyakutake approached the Sun from below Earth's orbital plane (grey sheet), then rose through it, passed closest to Earth, then raced onward to round the Sun and head outward again, all within a handful of weeks. The positions of Earth (blue) and the comet (red) are shown for various dates; the green line at May 1 marks the comet's closest approach to the Sun. (Nick James)

diffuse object. It lay right about where he had found his comet five weeks before. For an instant, he thought something was terribly wrong with the orbit of the Christmas comet. But then he remembered that comet had long since moved on, and besides he'd just glimpsed it over in a different part of the sky. With a jolt, Hyakutake realised he was seeing a *second* comet, placed by incredible coincidence in the same celestial region as his first one.

After he had walked around a little to calm down, Hyakutake sketched the new comet's position but could detect no movement, so once again he reported it as only a probable discovery. The new object appeared smaller and more condensed than the previous one. It was at about 11th magnitude in brightness and showed a round coma some 3 arc-minutes in diameter. When he checked its position in a star atlas at home,

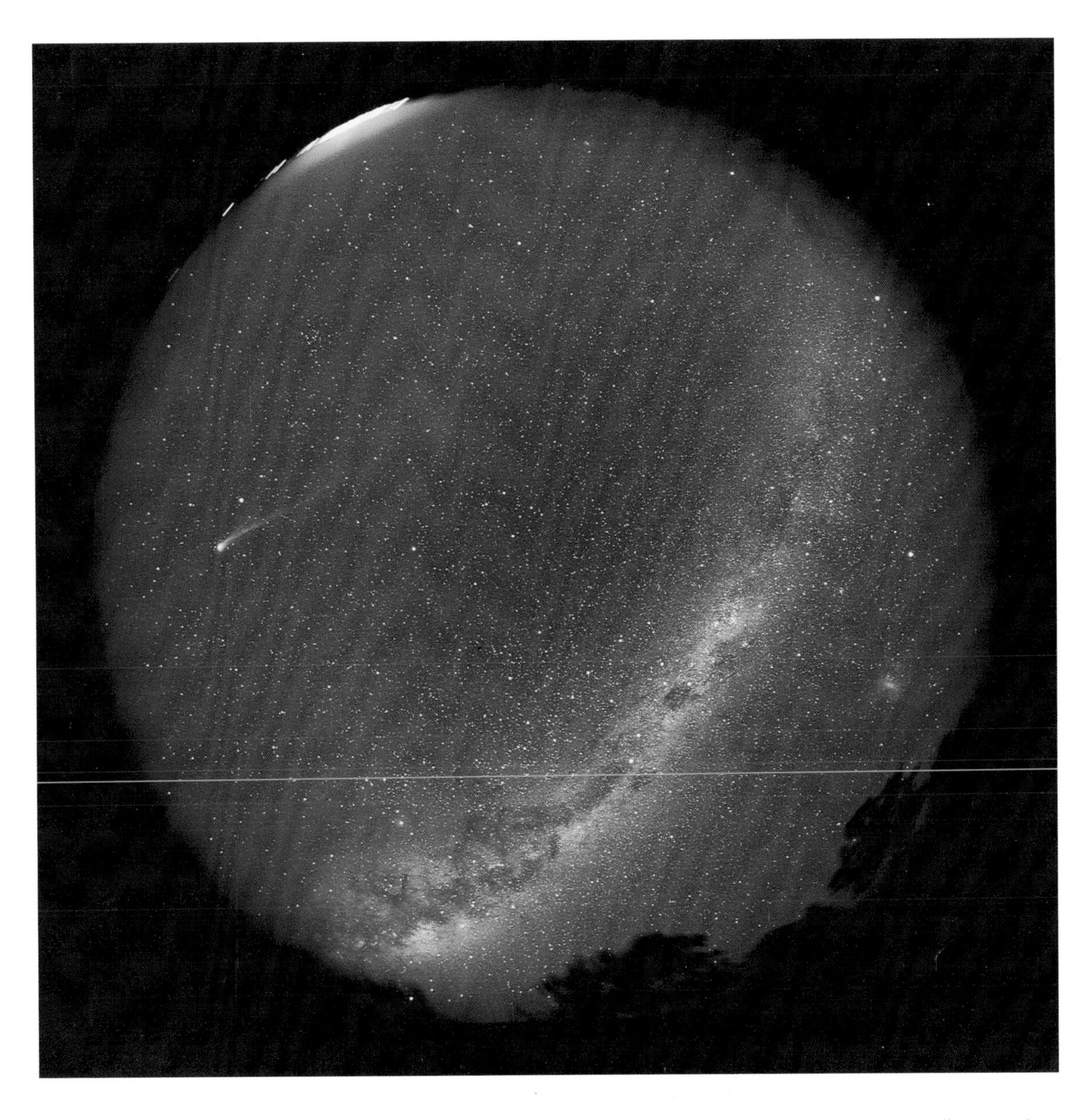

An all-sky view taken from Australia shows Hyakutake and its blue gas tail contrasting with the warmer tones of the star-rich Milky Way and the Large Magellanic Cloud (upper right). (Gordon Garradd; Loomberah, New South Wales; 16 mm f/3.5 fish-eye lens; hypered Fujicolor Super HGV 400; March 22, 1996)

By early March 1996, Hyakutake was becoming very active, as shown by this long exposure taken at the European Southern Observatory to study the structures in the comet's gas tail. (March 4, 1996)

Hyakutake realized that he had found the new object just 3° — about the width of a thumb held at arm's length — from the discovery site of his Christmas comet. A few days later, after confirmation by other observers in Japan and Australia, the Central Bureau announced the new discovery as comet C/1996 B2 (Hyakutake).

It wasn't long, though, before astronomers realized that this new Comet Hyakutake was going to behave very differently from the first.

Turbulence in the solar wind is churning Comet Hyakutake's gas tail in this composite image. (Kiso Observatory, University of Tokyo; March 23, 1996)

Calculations showed the new one's orbital path had last brought it through the inner solar system approximately 17,000 years ago. This figure is large and it contains a fair amount of uncertainty but, whatever the exact value, it is recent enough to show that Comet Hyakutake is a not a fresh comet straight from the Oort Cloud (see Chapter 1). These have orbital periods measured in millions of years. The meaning was clear: Comet Hyakutake's current orbit is not its original one, and its path has been shaped by at least a few encounters with the larger

Startlingly aqua in color, Hyakutake's coma shines with light from fluorescing diatomic carbon, ionized carbon monoxide, and sunlight reflecting from dust. In the blue gas tail, streamers betray where the solar wind's magnetic field lines are wrapping around the nucleus like tissue paper enfolding a prize apple. (Bill Whiddon; 8-inch f/1.5 Schmidt camera, 10 minute exposure on hypered Fuji HG 400; March 24, 1997)

planets. In other words, Comet Hyakutake has come this way before, but not very often.

More important, however, the new Comet Hyakutake's orbit was tailor-made to garner widespread attention: it had a track across the sky that would all but guarantee it becoming a Great Comet. The orbit would pass very close to Earth — 0.1 astronomical unit at nearest, or some 9 million miles. This meant the comet would appear bright. Moreover, the comet was going to traverse familiar and easy-to-find constellations. Third, the path skated close to the north celestial pole. This would make the comet visible all night long for viewers in the well-populated middle northern latitudes. And finally, the nights with the comet at its most visible would be moonless and dark.

Unlike Comet Hale-Bopp, which was discovered in July 1995 but didn't become prominent for another 20 months, the entire show for Great Comet Hyakutake was shoehorned into the first half of 1996. Of that period, the time that clinched the comet's reputation lasted a bare four weeks — roughly from March 15 to April 15 — when the comet was flying past Earth.

The brevity of best visibility resulted from both the comet's fairly small intrinsic size and the shape and orientation in space of its orbit, which meant that it spent only a short time close to Earth. Hyakutake's orbit is a skinny ellipse some 45 times longer than it is wide. (Think of an oval that's an inch wide and almost four feet long.) The inner end of the ellipse sits a toasty 0.23 AU away from the Sun, well inside the orbit of Mercury. And its outer end lies in the cold twilight more than 1,300 AU from the Sun, off in the direction of the southern constellation of Pyxis, the Mariner's Compass. Such an aphelion distance would place the comet well into the Kuiper Belt (see Chapter 1), except that the orbit makes an angle to the ecliptic, Earth's orbital plane, and lies almost wholly south of it. Thus when the comet is drifting slowly around its aphelion point, it probably lies outside the Kuiper Belt, which generally follows the plane of the ecliptic.

This description, however, applies only to the orbit that Hyakutake had as it approached us. As the comet passed through the planetary system its orbit underwent a major change. In essence, it got stretched.

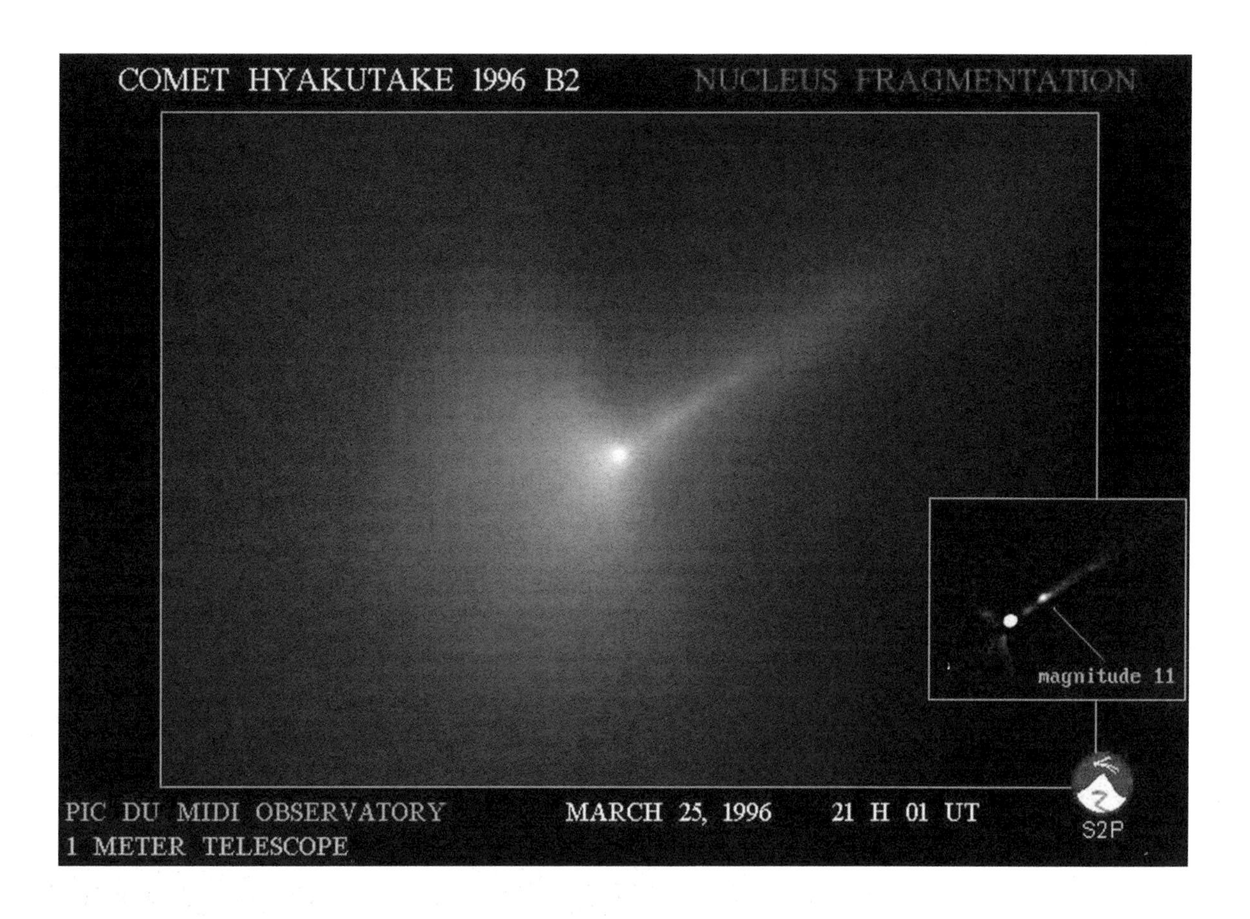

A dust fragment leaves the region of the nucleus in this image made by observers from the Bureau des Longitudes at France's Pic du Midi Observatory on March 25. (Observers: E. Frappa, J. Lecacheux, P. Laques, F. Colas, J. Klinger, A. Enzian, X. Cantorne)

Perihelion remains roughly where it was, but the aphelion point now lies almost 3,500 AU out, probably touching the Oort Cloud's inner regions. Increasing the aphelion distance also drastically changed the comet's orbital period: Yuji Hyakutake's Great Comet won't revisit us now for nearly 72,000 years.

At discovery, Comet Hyakutake lay 2.0 AU from the Sun (1.8 AU from Earth), and was approaching swiftly out of the southern half of the sky, brightening steadily as it came. On March 12, it crossed to the northern side of Earth's orbit, and two weeks later, on March 25, it raced past Earth at a minimum distance of 0.1018 AU, some 9.3 million miles (15 million km). This is about 40 times the distance to the Moon, and the close passage gave Hyakutake the distinction of being the nearest comet

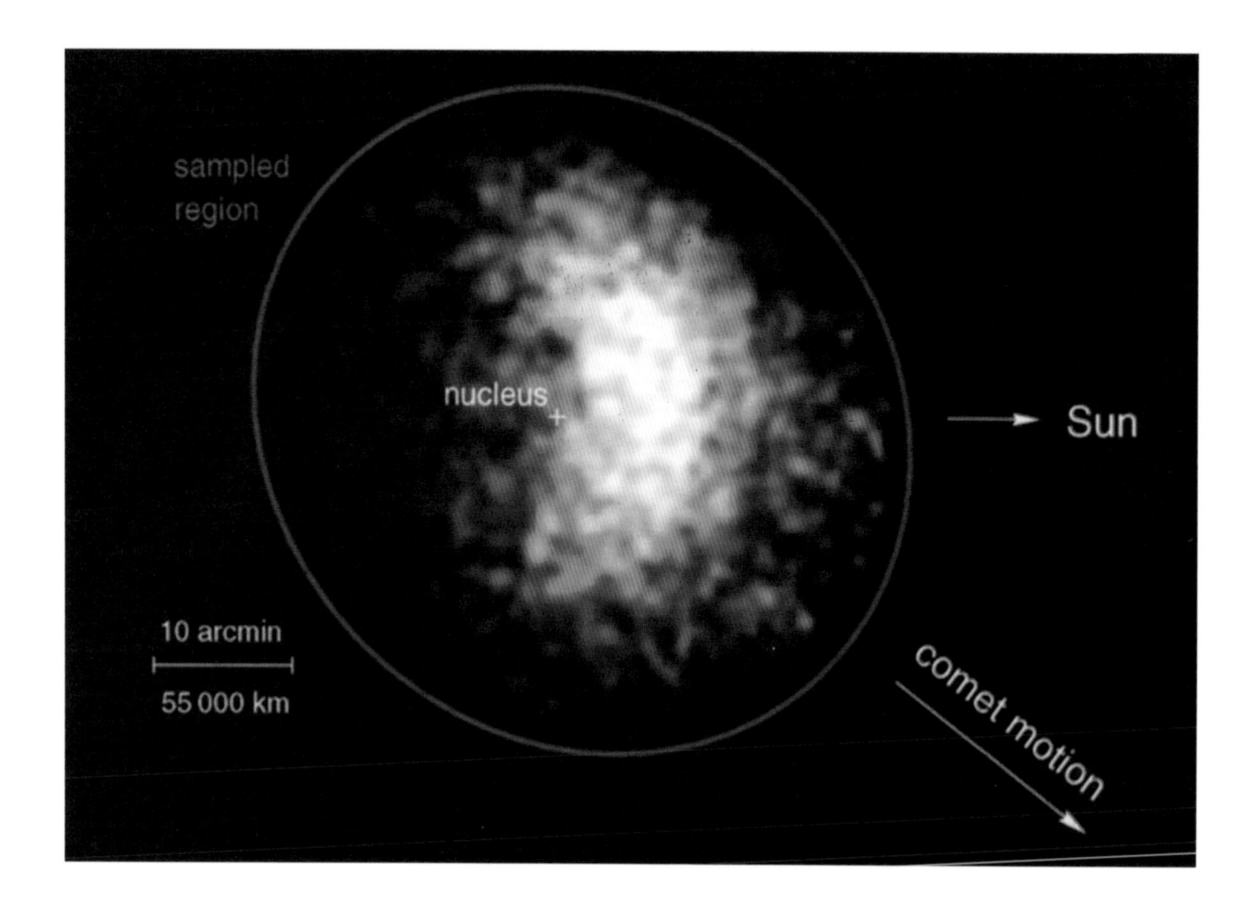

One of Comet Hyakutake's biggest surprises was the discovery that it is a source of X-rays. How the X-rays are produced remains unclear, but after the discovery with Hyakutake, scientists examined the records of other comets and found evidence of X-rays from them as well. (NASA/Goddard Space Flight Center; March 27, 1996)

since 1983 and the fifth nearest in the past 100 years. After skimming by Earth, the comet flew onward toward perihelion on May 1, 1996, where it rounded the Sun traveling 55 miles per second (88 km/s). Just five days later, it dove below the ecliptic heading outward, not to return again for a long, long time.

After the news of the comet's discovery and its grand prospects broke, observers with telescopes zeroed in on it. In early February 1996, during the cold hours before morning twilight began, the comet rose in the southeastern sky. It shone then at about 9th magnitude, and the coma had grown to 6 or 7 arc-minutes in diameter — roughly 300,000 miles (500,000 km) in diameter. The waning Moon moved into the region by the 8th and 9th, shedding a light that hampered observations until after the middle of the month when the phase was approaching New Moon.

As February progressed, Hyakutake brightened to become a binocular object, then a naked-eye one in the last few days of the month. The coma was now some 12 arc-minutes across and photos showed a 1° tail. By this time the comet was rising soon after midnight along with the stars of Libra, through which it wended a northward course, slowly at first but with night-to-night steps that grew steadily larger as the comet drew nearer to Earth.

As March began, the Moon was approaching Full. While its light hindered observers somewhat, the comet's growing brightness partially offset the lunar interference. Even when the waning Moon passed close to it on the 10th, the comet was still visible with binoculars. By mid-March, the comet appeared brighter each night, while photos showed a beautiful and turbulent gas tail. Around the third week of March, the broadcast media discovered the comet, and began to provide generally accurate information on how to see it. This greatly boosted the comet's worldwide audience. Also helpful was the fact that by then the comet was rising right after sunset in the eastern sky along with Arcturus and the other stars of Boötes. It was easy for anyone to walk outdoors on a clear evening and find the comet.

Moving swiftly northward on the sky across Draco and Ursa Minor, the comet brightened further as it went, reaching an estimated zero magnitude on the 24th. Telescopic observers were startled to note that the comet appeared to be breaking up, with a second false nucleus being visible deep in the coma. Further observations, however, showed that this was just a piece coming off the nucleus, not a total disruption of it.

High winds in space. Part-way back along the tail of Hyakutake is a knot of ionized gas caused by turbulence in the solar wind, which blows at hundreds of miles per second. Such knots are called disconnection events, and they help scientists trace the invisible gusts of protons and electrons that make up the solar wind. (Alessandro Dimai, Roberto Nuzzo, Francesco D'Arsiè, and Carlo Ferrigno; 300 mm f/2.8 lens, 15 minute exposure on Kodak Technical Pan film; March 25, 1996)

Meanwhile its tail outgrew even optimistic projections. On March 25th, the night the comet came closest to Earth, its gas tail reached a length of about 50°, more than a quarter of the way across the sky. Observers viewing under very dark skies reported tracing the tail for twice that distance. Japanese astronomers clocked a moving knot of turbulent gas in the ion tail that was racing away from the nucleus at almost 60 miles per second (100 km/s). The comet at that point was 9.3 million miles (15 million kilometers) from Earth, and the coma was estimated to shine at magnitude –0.8, right on the predictions. The comet's large head — some 1.5° across — was poised halfway between Polaris, the Pole Star, and the handle of the Big Dipper, while the ion tail traced its

A wider view of the March 25 disconnection event shows the knot in relation to Hyakutake's head and the main visible part of the gas tail. (Gary Seronik; 50 mm f/2 lens, 5 minutes on Kodak Gold 400; March 25, 1996)

Comet and galaxy. As the disconnection knot billowed away from the tail, it appeared to pass across the spiral galaxy M101. In reality, the galaxy lies over 20 million light-years from Earth. (Jerry Lodigruss, 400 mm f/2.8 lens, 10 minutes on hypered Fujicolor Super G 800 Plus; March 25, 1996)

glowing line across the handle. The next night it passed above Polaris, where its polar position kept the comet in view all night long.

Easy to see with the naked eye even from light-polluted cities, Great Comet Hyakutake made a wonderful sight for millions of people who ordinarily pay little attention to astronomy. And the media really swung into high gear, with newspapers and TV stations showing photos taken by local amateurs and professional astronomers and telling people how to go outside and spot the wonder in the sky. Even people without observing experience had no trouble finding it. All you had to do was go out and look toward the north.

A brief exposure (30 seconds) kept stars untrailed, as Tod Lauer photographed the comet above the dome of the 4-meter Mayall telescope at Kitt Peak. (ISO 1,000 film; March 25, 1996)

At this point, the comet was moving at a rate of nearly a degree per hour and telescopic viewers could see the comet move visibly in their eyepieces. This made photography tricky, as capturing sharp views of the comet meant that the camera had to track on the moving comet nucleus, whose motion differed from the ordinary rotation of the Earth. For any photo except those with very brief exposures, observers could get pinpoint stars and a blurred comet — or a sharp comet and trailed

The long tail of Hyakutake was its most memorable feature for many. At Joshua Tree National Park, California, Loke Kun Tan used a 6 × 7 cm format camera and a 165 mm f/2.8 lens to catch its blue gas tail. (April 13, 1996)

stars. But it proved all but impossible to capture both stars and comet sharply on those evenings.

Having passed closest to Earth, Comet Hyakutake now began to drop in brightness. The view, however, remained stunning, especially for telescope users. Many reported seeing a spike or jet coming from the central bright point in the coma. As the comet moved away from Polaris, it headed for Perseus, high in the northwest after sunset. And just beyond lay another lovely sight: the brilliant planet Venus near the Pleiades star cluster in Taurus. The comet's brightness continued to fall slowly. Confounding predictions (and disappointing observers), the comet did not begin to brighten again as it approached perihelion, due on May 1. Comets usually grow more active as they near the Sun, but much depends on whether the active areas on the nucleus continue to "see" the sunlight. As the comet moves through perihelion, the Sun's location in the sky over the nucleus will change. And if the active areas do not remain strongly sunlit — or experience solar heating for a smaller part of the nucleus' day — they will slow their activity. Perhaps this is what happened with Hyakutake.

In any case, when the calendar pages flicked over into April, a small

Toward the middle of April, the comet was sinking in the west in the evening sky. John Volk photographed it against the stars of Perseus, with blazing-bright Venus at upper left and the Pleiades star cluster in the trees, center left. (6 × 7 cm format, 105 mm f/3.0 lens, 8 minutes on hypered Konica Super XG400; April 13, 1996)

dust tail about 5° long began to emerge to accompany the gas tail, which extended several times longer. (The visible length of both depended a lot on how dark a site one viewed from.) As moonlight increased in the first days of April, the comet receded from Earth. On the evening of April 3 to 4, a total lunar eclipse darkened the Full Moon for an hour and a half.

While April continued, the comet dropped steadily in brightness, although it remained impressive in the northwestern sky after sunset. On the 19th and 20th, Mercury posed with Hyakutake, a thin crescent Moon, and the much higher Venus to make one last striking scene. Although heading steadily closer to the Sun, the comet's activity and brightness remained stuck at about magnitude 2.5 — fairly dim for an object now sinking quite low in a twilight sky. For most members of the public, the Great Comet Show of 1996 was over — even though the comet had yet to pass perihelion, which duly came on May 1.

As it skimmed over the Sun and was lost in the glare for viewers on Earth, Comet Hyakutake was photographed by a Sun-monitoring satellite, the Solar and Heliospheric Observatory (SOHO) on May 1, 2, and 3. SOHO's images showed the comet and three tails against a background of the Sun's million-degree-hot inner corona. The three distinct tails arose from the differing weights and sizes of particles ejected from the nucleus. The heaviest dust particles left the nucleus at slower speeds and remained closer to it as it tore around the perihelic hairpin turn, while lighter dust was driven straight away from the Sun by radiation pressure. Lastly, the solar wind carried off atomic particles from the comet and aligned them with the corona's magnetic field.

The comet's post-perihelion course took it southward on the sky at great speed. After it emerged from the Sun's blaze of light, the comet's path strongly favored Southern Hemisphere observers, who picked it up in the morning twilight on May 9. But its brightness was dropping for them too. While Hyakutake hung on as a naked-eye-visible object until the end of June, many southern observers were forced to use telescopes and binoculars for decent views. The coma now measured about 6 arcminutes across, while the tail stretched a few degrees. Over July and August, the comet raced away from the Sun and Earth, heading back into depths of space. As sunlight gave less and less warmth to the comet, its activity dwindled along with its brightness: 7th magnitude on

July 1, 8th magnitude August 1, 11th magnitude September 1. When last photographed late in October 1996, Great Comet Hyakutake was at 17th magnitude and displayed no coma — only a star-like point drifting slowly against the stellar background.

For scientists Great Comet Hyakutake's arrival was a bonanza, although its riches weren't extracted without cost, mainly in observing plans that had to be improvised in great haste. The comet's rapid apparent motion also nearly reached the limit of what large professional telescopes could track on. Moreover, the short advance warning meant that the comet's path in the sky could not be nailed down with complete accuracy until right before an observation was to be made. This snag didn't matter to anyone just looking at the comet or even photographing it with backyard instruments. But professional telescopes need precise pointing information to work effectively — and this goes double for telescopes working at radio and far-infrared wavelengths. Because observers cannot look through these instruments and simply point them the way they can an optical telescope, they have to aim the telescopes blindly — yet with great precision — according to where the comet is predicted to be.

As Hyakutake approached the Sun, observers tracked the growing cloud of water vapor that surrounded it, noting that Comet Hyakutake seemed to be generally producing water about as fast as Comet Halley did at similar distances from the Sun. Observers did note, however, that at perihelion some water molecules were flying out of icy grains in the coma at unheard-of speeds, 14 to 28 miles per second (22 to 44 km/s). The reasons for the high ejection speed of the water vapor are not yet known, although violent activity in the Sun's corona which was going on at the same time could play a role. (There's a lot still to be learned about what happens when you put a snowball next to the Sun!)

Dust particle sizes generally matched Halley's, although Hyakutake was spewing much more dust overall — up to 10 tons per second by mid-April, blasted off at 1,000 miles an hour (500 m/s). These findings led to initial estimates that Hyakutake's nucleus was about the same size as Halley's. (Radar results later downgraded its size drastically.) Along with water vapor, a host of other substances came boiling out of the ice

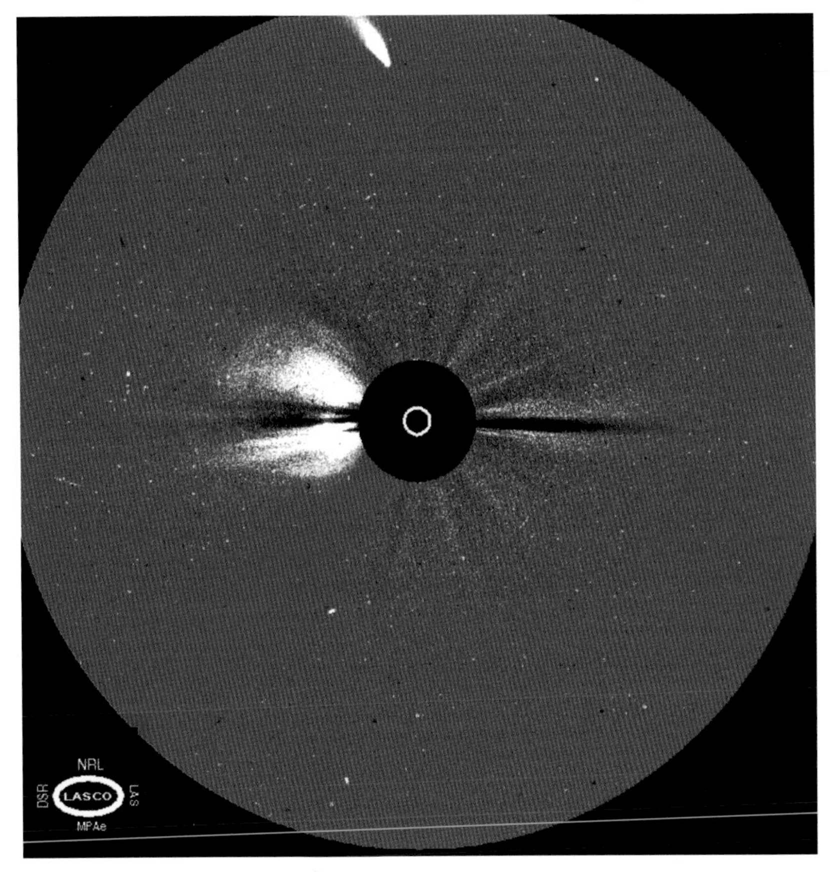

Hairpin turn. As Comet Hyakutake sped through perihelion on May 1, it was caught by the coronal imager of the Solar and Heliospheric Observatory satellite (SOHO). By coincidence, SOHO's instrument also captured a huge outburst in the corona that was going on at the same time (left). (NASA/Goddard Space Flight Center; May 1, 1996)

and dust of the comet's nucleus. These included carbon (soot, essentially), carbon monoxide (CO), carbon dioxide (CO_2), methanol (CH_3OH), cyanogen (CN), hydrogen cyanide (HCN), hydrogen isocyanide (HNC), methyl cyanide (CH_3CN), formaldehyde (H_2CO), hydrogen sulfide (H_2S), ammonia (NH_3), and molecular sulfur (S_2).

The head of Hyakutake showed a strong turquoise or aquamarine color, easily noticable even to the unaided eye. This came largely from diatomic carbon (C_2) emission, perhaps mixed with a warmer tint from sunlight reflecting off dust. The bluish color of the tail, also visible to the eye and captured in many photos, came mainly from ionized carbon monoxide, a common ingredient of comets. The tail itself was long and beautiful, displaying changes as it twisted and turned in the gusts of the solar wind. On February 28 and March 10 and 25, observers noted disconnection events, where turbulence in the solar wind snipped off part of the gas tail.

While the carbon monoxide was expected, the comet handed scientists two big surprises. The first was the discovery of abundant methane (CH_4), acetylene (C_2H_2), and ethane (C_2H_6) in it. These hydrocarbon substances had never been observed in comets before. The ethane and methane in particular were surprising, and they hint that Hyakutake may contain surviving interstellar material — ices that condensed onto interstellar dust grains before the solar nebula came together and somehow stayed intact inside the comet as it formed. The ethane may be the result of sunlight-driven chemical reactions in the coma, and it might also result from low-temperature reactions within the comet during the ages when it orbits far from the Sun. In any case, further study will help scientists better understand where this comet formed, and how its pieces came together.

The second surprise was X-ray emission, another cometary first. Detected by the German satellite observatory ROSAT, the X-rays came from the sunward side of the coma and flickered in strength over a few hours. After much study of the data and a lot of argument over the meaning, scientists think the emission came mainly from the interaction of charged particles in the solar wind with the comet's atoms and molecules. They also think some X-ray emission could also come from dust particles colliding at extreme velocities.

Hyakutake's close flyby gave observers using the 1,000-foot Arecibo radio telescope in Puerto Rico a chance to bounce radar beams off the comet's unseen nucleus, probing its size and nature. Rather to their surprise, they found that it was less than 2 miles (under 3 km) in diameter — small compared with Halley's nucleus and tiny next to that of Hale-Bopp. This meant that Hyakutake was intrinsically a far more active comet than Halley. Radar showed the solid nucleus was surrounded by a blizzard of rubble and fluffy "rocks" half an inch (a centimeter) and larger in size whose pieces were flying off the nucleus at about 30 miles an hour (12 m/s). The echo from the cloud of particles was ten times stronger than from the nucleus itself, testimony to their numbers. The particle cloud helped explain other observations which indicated that most of the water vapor and other gases were coming from a larger, more widespread source than just the nucleus alone.

Many observers with telescopes (amateur and professional) reported seeing spiral fountains of light in the inner coma. These come from jets on the sunward side of the nucleus, and reveal that the nucleus was rotating once every 6.23 hours (6h 13m), extremely fast by comparison with Halley, whose nucleus has a complicated rotation period measured in days. This may also have helped create the "boulder storm" as pieces that worked loose were shrugged off by the spinning nucleus.

In short, while the nucleus of Hyakutake was smaller than the nucleus of Halley, it was far more active. This makes sense, given that Halley surely has made many more trips around the Sun than Hyakutake has. It's entirely reasonable that a thicker mantle of dust would cover Halley's nucleus compared with Hyakutake's, which appears to remain in a relatively youthful and active state.

As Great Comet Hyakutake faded, planetary scientists took stock of their data and began to fit them into an evolving picture of this comet and others. Meanwhile, they were also looking over their shoulders. Another Great Comet was on the way — Comet Hale-Bopp — and it promised to be as much of a revelation as Hyakutake.

4 *Great Comet Hale-Bopp (1997)*

The night it was discovered, Comet Hale-Bopp looked anything but "great" to the eye. When Alan Hale and Thomas Bopp first spotted the comet on July 23, 1995, it was just a small hazy patch of light shining dimly at 11th magnitude, more than 100 times too faint to be seen by the unaided eye. It resembled a globular star cluster, one of the many that litter the Milky Way skies of Sagittarius, the Archer, the constellation where the comet appeared.

Alan Hale was at home in Cloudcroft, New Mexico, a small town high in the Sacramento mountains. In his forties, Hale is a comet scientist by profession, and an amateur comet-hunter with more than 400 hours' search-time in his logbook. On this particular evening, however, he was simply making brightness estimates of known comets. Observing from his driveway with a 16-inch Newtonian reflector telescope, he was waiting for a particular comet to rise so he could observe it. Since the night was the first clear one in weeks and he had an hour to kill, Hale decided to skygaze in Sagittarius, home to many beautiful star clusters and glowing clouds of gas.

Down in the southern part of the constellation, he came upon the globular star cluster M70, which looked like a soft-edged tennis ball. Just to its east lay a second fuzzy object, much smaller and dimmer but generally resembling M70. Hale didn't think that M70 had a companion cluster like that, so he checked a star atlas. What he found there — that no other object was plotted next to M70 — supported his sudden hunch that he might have a comet in the eyepiece. Hale determined the mysterious object's position and soon verified that it was moving, the unmistakable hallmark of a comet. Further checks using the Internet ruled out the possibility of it being already known, and Hale excitedly reported the new find. It was his first comet discovery.

That same night a group of amateur astronomers from Phoenix, Arizona, had trekked into the desert 90 miles south of the city with their telescopes. Comet-hunting lay far from their plans. They were bent on exploring star clusters and nebulae unhindered by the skyglow from city lights. Among the group was Thomas Bopp, a keen deep-sky observer. Bopp, who is also in his forties, didn't even own a telescope but made his observations using telescopes borrowed from his friends.

Not long after Hale made his cruise through southern Sagittarius, Bopp and his friend Jim Stevens reached the vicinity of M70 also. Stevens owned the 17.5-inch Newtonian reflector that he and Bopp were using, although Bopp was at the eyepiece when the fuzzy image of the comet drifted into view near the globular cluster. Like Hale, Bopp was somewhat puzzled at the sight and so he checked a star atlas — only to find, just as Hale had, that M70 stood alone in its particular piece of sky. After observing the object long enough to be convinced it was a comet and to note a position for it, Bopp raced back into town to make his report of it. As in the case with Hale, the comet was Tom Bopp's first.

The newly discovered comet, officially dubbed Comet Hale-Bopp (1995 O1), soon showed surprising qualities that hinted that it could turn out to be a Great Comet. To begin with, Comet Hale-Bopp was found at an astounding 7.2 astronomical units (AU) from the Sun, a distance that would place it about halfway between the orbits of Jupiter and Saturn. (Earth orbits the Sun at 1 AU.) Comet Hale-Bopp was 666 million miles (1.1 billion km) from the Sun; most comets are discovered at about half that distance, and none had ever been found so far out.

A second surprise was the new comet's relative brightness (11th magnitude), and the fact that it already showed a hazy coma. Developing a coma so far from the Sun proclaimed that Hale-Bopp was far more active than ordinary comets. At 7.2 AU nearly all comets are inert chunks of dirty ice dimly reflecting the weak sunlight that reaches them. They look purely star-like and only their slow movement betrays their non-stellar nature. Yet here was Comet Hale-Bopp, fizzing along in a cloud of dusty gas, with its ices cooking in the Sun's feeble rays.

What drove such activity? Observations made with professional telescopes and detectors showed that, while the coma was made mostly of dust, it was permeated by a haze of carbon monoxide (CO) and cyanogen (CN) gas. These two volatile substances had boiled out of the ices making up the comet's nucleus, blowing off huge quantities of dust in the process. In all, the coma extended about 2 million miles (3 million km)

The diminutive star cluster M34 lay close to the head of Hale-Bopp on April 5. Note the streamers in the blue gas tail, produced as the solar wind's magnetic field lines wrap around the comet's nucleus. (Ben Gendre, 5.5-inch Schmidt camera, 5 minutes on Kodak Royal Gold 100)

in diameter, more than twice the diameter of the Sun. Yet it was so delicate that the pressure of sunlight was pushing this ball of dusty gas slightly off-center from the nucleus. Soon after the discovery, one comet scientist suggested that the offset coma, which also showed a spiral structure, could have come from an outburst in one area on the comet's nucleus shortly before it was found. If true, the outburst may have contributed to the comet's discovery by making it brighter and more visible. Other scientists dispute this and the question remains unresolved. But no matter what happened around the time of discovery, Comet Hale-Bopp was so rich in volatile ices that its dusty coma had probably become well established four or five years before it was found by Hale and Bopp. At that time the comet would have lain at a distance of 18 to 20 AU — about as far from the Sun as the orbit of Uranus. This also emphasized how unusually active Comet Hale-Bopp was, since activity at Uranus' distance is quite rare.

In any event, whether or not Comet Hale-Bopp was abnormally bright that July night is beside the point, really. This comet was ripe for finding. Sagittarius is a popular region for skygazing during June, July, and August. If Alan Hale and Tom Bopp hadn't caught the comet when they did, it would have been picked up two nights later when another amateur astronomer made his own independent discovery of it. Comet Hale-Bopp was simply passing through the right piece of sky at the right time of year.

As new observations of the comet's position streamed in — and with the finding of a lucky prediscovery photo taken in April 1993 — scientists calculated an orbit for the new body. Comet Hale-Bopp travels in a much-elongated oval tilted almost at right angles to the ecliptic, the plane of Earth's orbit. (The ecliptic is where most of the other planets orbit or very nearly.) If you envision the ecliptic as the flat top of a table with Earth's orbit scribing a circle on it, Comet Hale-Bopp's path on its approach to the Sun came from down near the floor and rose up through the table top at an angle of about 45°. Soon after its discovery, orbital calculations showed that Hale-Bopp would fly past Earth at a distance of 1.3 AU on March 22, 1997 and swing round the Sun (the perihelion point) on April 1 at a distance of 0.9 AU. About a month later, on May 6, Hale-Bopp would dive down through the ecliptic (the table top) again at

Just a smudge of light — Comet Hale-Bopp, five weeks after discovery. The comet was more than seven times farther from the Sun than Earth, but already it was brighter than any other comet at that distance. Photo taken September 1, 1995, with the University of Hawaii's 2.2-meter telescope. (David Tholen & Richard Wainscoat/University of Hawaii–IFA)

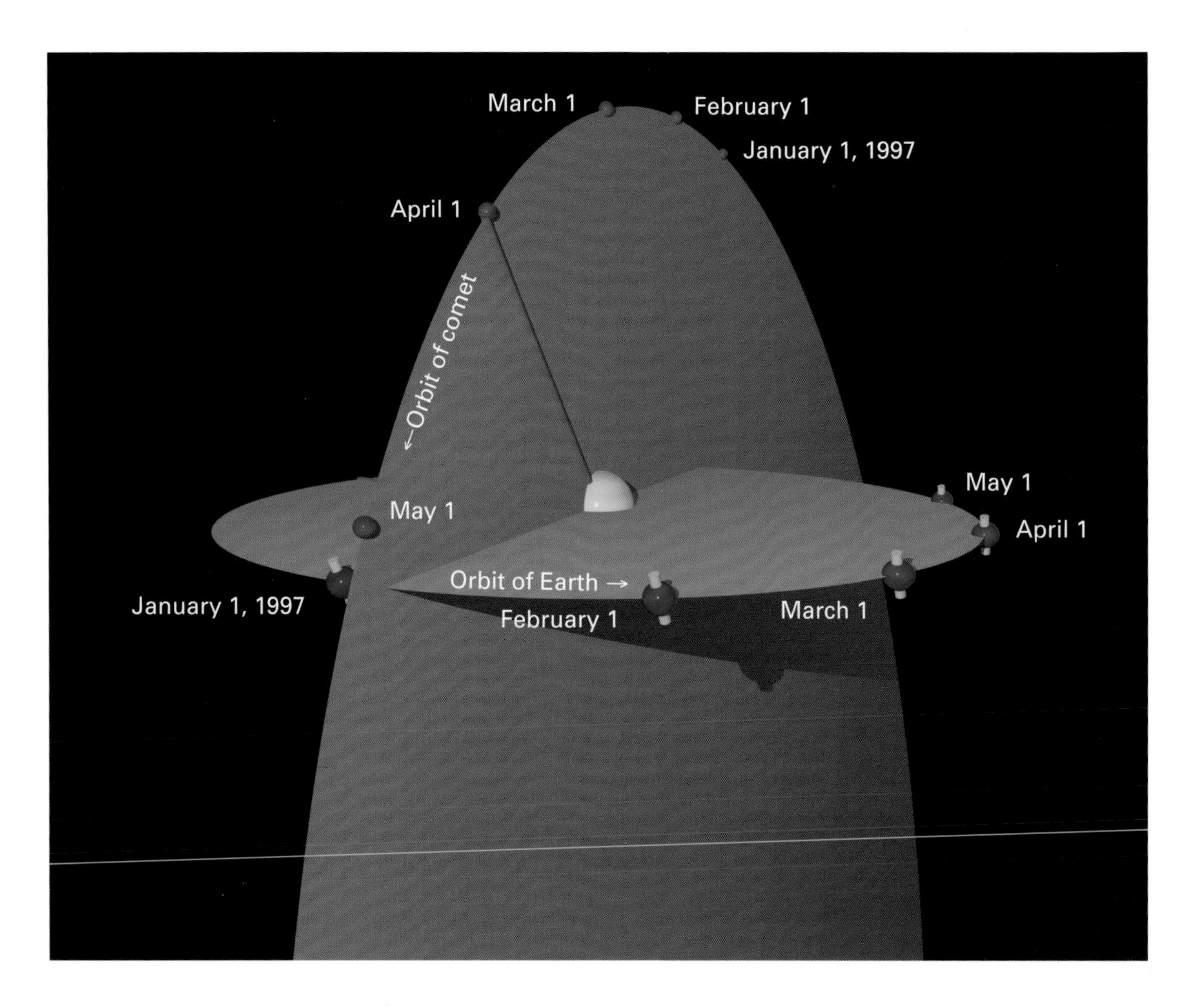

The orbital plane of Comet Hale-Bopp makes almost at a right angle to the plane of Earth's orbit around the Sun (grey surface). This helped keep the comet well-separated visually from the Sun during most of the visit. The positions of Earth (blue) and the comet (red) are shown for the first day of each month; the green line for April 1 marks the comet's closest approach to the Sun. (Nick James)

the start of its long trek out to aphelion, its farthest distance from the Sun. In all, the comet spent a little over 14 months "above" the ecliptic (from late February 1996 to early May 1997), during which time the comet came inside the orbit of Jupiter and Mars, passed Earth, made its turn around the Sun, and headed outward again toward aphelion.

The aphelion that Hale-Bopp is heading for this time is a lot nearer the Sun than its previous one, thanks to an encounter with Jupiter that occurred on the comet's inbound course. On April 5, 1996, while Hale-Bopp was still approaching the Sun, it passed 72 million miles (116 million km) away from Jupiter, relatively close in astronomical terms. Jupiter's strong gravity deflected Hale-Bopp's course a little, most notably reducing the comet's aphelion distance from some 525 AU

Evolution of a nucleus. The activity in and around Hale-Bopp's nucleus shows changes over the 13 months covered in this Hubble Space Telescope survey. Note especially the dust outburst in September 1995, with its curving bar, produced when the spinning nucleus turned away from the debris it had ejected. (Harold Weaver, Johns Hopkins University; Space Telescope Science Institute; NASA)

to about 358 AU. (The new aphelion still lies about nine times farther from the Sun than Pluto's orbital distance of 40 AU.) The change also shortened the comet's orbital period from 4,265 years to 2,404 years. Hale-Bopp was last seen around 2268 BC, it will reach aphelion around AD 3199 (in the southern constellation of Pavo, the Peacock), and will be back in our neighborhood again early in the forty-fifth century, around the year AD 4401.

Hale-Bopp's current orbit is an ellipse like Hyakutake's, but rounder in shape and it doesn't reach as deep into space. Whereas Hyakutake's new orbital ellipse stretches out to more than a thousand astronomical units and is about 45 times longer than it is wide, Hale-Bopp's new orbit extends only 10 times its width, and its aphelion remains within a few hundred astronomical units of the Sun.

Orbital changes are nothing new for Comet Hale-Bopp, however. In the jargon of celestial mechanics, the comet's orbit is termed chaotic. This means it is highly susceptible to drastic alteration by the gravitational tugs of the planets, chiefly Jupiter. A chaotic orbit poses enor-

The eye recorded a somewhat different comet than film did. This sketch of Hale-Bopp was made from memory by Gregg Geist on February 15, 1997, after observing with a 6-inch Dobsonian reflector telescope and a low-power eyepiece.

A windmill near Payson, Arizona, made a good foreground for John Williams, who caught Comet Hale-Bopp on March 3,1997. He used a 50 mm lens at f/2 with a 20 second exposure on Fuji ISO 1,600 print film.

mous difficulties for anyone trying to calculate the comet's long-term past or future course. Despite the uncertainties, we can paint a general picture of Hale-Bopp's probable origin. The comet likely formed some 4.5 billion years ago in the outer solar system, perhaps near the present orbit of Uranus. From there it was perturbed repeatedly by multiple planetary encounters, mostly through the action of Neptune and Jupiter. The encounters eventually reshaped its orbit into the path the comet arrived on this time. A less likely possibility is that Hale-Bopp formed farther away, in the Kuiper Belt, and was kicked by Neptune out to the distant Oort Cloud, whence it returned thanks to the gravitational effect of a passing star or a galactic cloud of dust.

The reason scientists think that Hale-Bopp was born near Uranus instead of farther out comes from its chemical composition. Studies showed that Hale-Bopp is deficient in neon, a substance found only in comets that form under extreme cold. Thus Hale-Bopp was probably born in slightly warmer regions nearer to the Sun than the Kuiper Belt. (Its relatively low proportion of carbon monoxide also supports this notion.) Uranus' distance could be just about right.

Hale-Bopp has circled the Sun many times already, but it is still comparatively fresh. This means it is not as seasoned a comet as Halley's, for example, which has made a great many returns. (So far, 30 consecutive passages for Comet Halley have been identified in historical records.) Also supporting the notion that Hale-Bopp is a relative newcomer is its unusually strong activity, which produced, area for area, over 100 times more dust than Halley.

The eventual fate of Comet Hale-Bopp is something scientists can only guess at. But the alternatives are all pretty stark: most objects in chaotic orbits eventually hit something, either the Sun or a planet. A few chaotic comets are thrown out of the system to the Oort Cloud or to interstellar space, and even fewer end up in orbits that placidly circle the Sun outside the realm of the planets, never again approaching the Sun closely enough to warm up and become active once more. But no one can say for sure just which fate lies in store for Hale-Bopp.

Alan Hale and Tom Bopp discovered their namesake comet in the southern Milky Way constellation of Sagittarius. As the comet approached the Sun more or less in a straight line, observers watched it trace a path that ran northward against the stellar backdrop of the Milky Way. While the comet ran relatively straight, its course among the stars described broad leisurely zigzags. These reflected our moving terrestrial viewpoint that shifts back and forth as we ride Earth's yearly circuit around the Sun like a merry-go-round. Because of the comet's great initial distance, it remained within Sagittarius for the entire year following its discovery. Then, moving more quickly as it drew nearer, Hale-Bopp cruised across the constellations of Scutum and Serpens Cauda into

Blue gas and white dust tails stand out in this photo of Comet Hale-Bopp from March 7, 1997, when the comet was passing from Cygnus into Lacerta. (Loke Kun Tan; Celestron/Epoch 8-inch f/1.5 Schmidt camera, 8 minutes on unhypered Kodak Pro 100 PRN)

OPPOSITE The ruddy glow of the North America Nebula in Cygnus shows clearly amidst the stars of the Milky Way, as Hale-Bopp sails along to the northeast. (Joe Orman; 50mm f/1.8 lens, 4 minutes exposure on Ektachrome P1600; March 9, 1997)

Shells of dust dominate the inner portion of the coma around the nucleus of Hale-Bopp, while turbulent, finely structured rays weave through the plasma envelope farther out. This ultraviolet image was taken on March 15 at the Zeiss 2-meter Telescope operated on Pik Terskol, Caucasus Mountains, by the International Center for Astronomical, Medical, and Ecological Studies, Kiev. (Klaus Jockers, Till Credner, Tanyu Bonev, Max-Planck-Institut für Aeronomie; March 15, 1997)

Ophiuchus, only to turn again and cross a more northerly part of Serpens, reaching Aquila in January 1997.

Its path took flight as the comet approached Earth's vicinity. Between January and March 1997, it sped along the Milky Way, growing steadily brighter and more magnificent as it passed through Sagitta, Vulpecula, and Cygnus in rapid succession. Carving a course eastward across Lacerta, the comet reached its farthest northern extent — and greatest visibility — in Andromeda during late March 1997. It still lay in Andromeda when it passed its point of closest approach to the Sun on April 1. Then heading outward again and fading very slowly at first, it drove southeast into Perseus. Next it cut across Taurus and, fading more quickly now, it swung down through Orion, Monoceros, Canis Major, and Puppis. Hale-Bopp finished 1997 in the southern constellation of Dorado, not far from the bright star Canopus in Carina. During 1998, the comet drew a lazy, open oval path in the vicinity of Canopus. By early in 2000, it will lie in Mensa.

For amateur and professional astronomers, the whole apparition lasted nearly two years — unprecedentedly long for a comet — and was filled with highlights. Observers started following the comet as soon as it was discovered in July 1995 and they continued to track it as Sagittarius slipped into the twilight glow of the Sun that autumn. Between mid-August and late October 1995, comet watchers noted five outbursts occurring roughly every 19 days. Each outburst brightened the comet by 2 or 3 magnitudes, about 5 to 15 times, before it dimmed again in a day or so. Typically, an outburst of dust was followed by the development of luminous jets and shells, which slowly faded into the background of the coma. These repeated explosions of dust were driven by eruptions of gas that attained speeds of 150 miles per hour (240 km/h). Changes in the outbursts also hinted that the nucleus was rotating in a complex way — a mix of spinning and tumbling — but the results were unclear. By the end of the apparition, after many months of study, scientists finally decided that Hale-Bopp's day lasts 11.34 hours (11h 20m). There were hints that the comet might also be rotating around a second axis, but more analysis will be needed to settle the question.

Observations that noted the huge quantities of dust released suggested that Hale-Bopp's nucleus was unusually large. Early estimates

David Feltham caught Hale-Bopp over the spire of Stockholm's City Hall. He used a 28–105 mm zoom lens set at about 50 mm. The exposure was roughly 4 seconds at f/4 on Kodacolor Gold 200 film (March 19, 1997)

In Canada, Eric Hayes photographed the comet above a farmhouse in Cookville, Nova Scotia. His camera had an 85 mm lends and Hayes used a 15-second exposure at f/2.8 on Fujicolor 800 film (March 20, 1997)

pointed to a diameter for the nucleus of about 25 miles (40 km). This was several times the size of Halley's nucleus (10 miles) and more than ten times bigger than Hyakutake's (2 miles). The data also suggested that Hale-Bopp's nucleus might be elongated, which would be no surprise since even a 25-mile diameter is far too small for the nucleus to squash itself round through self-gravity. Besides, the very nature of a comet nucleus makes it inevitable that deep hollows will erode where jets are active. While the 25-mile diameter underwent continual refinement over nearly two years of study, in the end the figure fell just about in the middle of the range observers could agree on. Hale-Bopp really was a big comet.

The last observations of 1995 came in late November, after which the comet was too close to the Sun to be seen. In February 1996, when observers were recovering Hale-Bopp as it emerged from the Sun's glare

Till Credner was at the Pik Terskol Observatory in Russia when he photographed Hale-Bopp over Mt. Elbruz, the highest peak in the northern Caucasus Mountains. He used Fuji Super G 800 film, a 35 mm lens, and a 15-second exposure at f/1.4 (March 30, 1997)

low in the morning twilight, word began to spread about a new comet just discovered in Japan: Comet Hyakutake (1996 B2). This new comet brightened rapidly and began the breathtaking dash that took it across the northern skies in a blaze of glory (see the previous chapter). It made a beautiful forerunner for Hale-Bopp, and all the more wonderful for being so unexpected.

During the early months of 1996, Hale-Bopp emerged from the

The domes of Yerkes Observatory in Wisconsin gleamed in the light of a nearly full Moon when Dean Armstrong photographed Comet Hale-Bopp over the observatory. He used Fuji Super G 800 film and a 50 mm lens set to f/2.8; the exposure lasted 15 seconds (March 23, 1997)

predawn twilight and brightened steadily as it moved northward generally along the Milky Way. By the end of March, the comet shone at 8th magnitude — still too faint for the unaided eye, but easy in binoculars and a fine sight in a telescope. As its orbit brought it closer to the Sun, Hale-Bopp felt an increasing warmth and its activity grew. Observers studied its coma and burgeoning tail, the latter reaching about half a degree in length. To everyone's relief, the comet continued to brighten as predicted, lending confidence that it would indeed make a spectacular sight a year hence, in March and April of 1997. (No Comet Kohoutek here!)

As summer 1996 began, Comet Hale-Bopp brightened to the point where experienced observers viewing from a dark-sky site could just

Pale red from an aurora tinged the sky near the comet for Darren Talbot, observing on March 28, 1997, from Beaverbank, Nova Scotia, Canada. He photographed the scene with a 28 mm f/2.8 lens and a 2-minute exposure on Kodak Royal Gold 1,000 film.

For a couple of nights at the end of March 1997, Hale-Bopp glided by the Great Galaxy in Andromeda (M31). (Alessandro Dimai, Renzo Volcan, Alessandro Zardini; 100 mm f/2.8 lens, 3 minutes on hypered Kodak Pro Gold 400 II; March 27, 1997)

The multiple antennas of the French–German–Spanish IRAM millimeter-wave radio telescope make an eerie foreground for the comet. (Andre Rambaud; 28 mm f/2.8 lens, 45 second exposure on Kodak Ektar ISO 1,000; March 31, 1997)

detect it by eye alone, without using a telescope or binoculars. The comet was now about as far from the Sun as the outer part of the asteroid belt. By autumn 1996, the comet had sailed inside of 3 AU, which meant that sunlight was strong enough to vigorously evaporate the water ice in its nucleus. This spurred the comet's activity to new heights and brightened it further. The coma was now about half a degree across, and the dust tail reached twice that. Also around this time, photographers were recording as many as seven jets radiating out of the inner coma, like the arms of a not-quite-complete spider or octopus. One jet reached a length of at least 19,000 miles (30,000 km).

On September 26, 1996, a total lunar eclipse (visible from western Europe and through the Americas) dimmed the Full Moon for over an hour. With moonlight much reduced, observers could trace the comet's tail, by now some 2° long. By November, when the comet was once again slipping into evening twilight, it was bright enough for excellent views by telescope and good ones by eye, although both were hampered somewhat due to the comet's low elevation and the glow of twilight.

After passing through the Sun's glare for the second time, Hale-Bopp returned to center stage early in 1997 for its final and climactic act. It became visible in the predawn sky in January 1997, low in the east among the stars of Aquila. At this point it lay about as far from the Sun as the orbit of Mars, or 1.5 AU. Its tail did not look like much at first because it was extending almost directly away from Earth. During January and February, the comet continued its northward path along the Milky Way, and rapidly began to rival the prominence of Comet Hyakutake almost a year before. Hale-Bopp now shone at 2nd magnitude. The passing weeks altered the viewing geometry so that the comet revealed a dust tail that now stretched some 5° to 7°. One observatory reported photographing a kink in the gas tail, probably caused when the comet crossed a magnetic field boundary in the solar wind. In the coma, complex structures of dust shells, jets, and hoods appeared and slowly changed shape, as the invisible nucleus whirled and spun twice a day in the warm sunlight, erupting in ever-renewed fountains of dusty gas.

Early March 1997 marked when the comet started to break into the general public's consciousness, thanks to radio and TV coverage. Yet, because Hale-Bopp was still best visible in the icy skies of dawn, only the most avid comet-watchers were outdoors observing it faithfully every clear morning. Day by day, the comet moved northeasterly from Cygnus through Lacerta and into Andromeda. (On March 9, a total solar eclipse over parts of Mongolia and eastern Siberia let observers in the path of totality glimpse the comet in daytime. But the views were disappointing because the sky remained fairly bright even at full eclipse.). Around this time, the comet had moved far enough north in the sky that it was visible

The twelfth-century towers of a ruined church near Herne Bay, Kent, England, point to Hale-Bopp. (Paul Sutherland; 24 mm f/2.8 lens, Scotchchrome 800/3,200 processed at ISO 1,600; March 31, 1997)

all night long, especially for observers above 45° north latitude. It shone at about zero magnitude.

Then as the comet cruised eastward just south of the Milky Way, it finally became easier to watch in the evening sky than in the predawn. This, along with its increasing brightness (which reached about –0.5

Caught between the two tails, the Double Cluster in Perseus makes a fine contrast with Hale-Bopp. Near the comet's head lies another open star cluster, M34. (Loke Kun Tan; 6 × 7 cm format, 165 mm f/2.8 lens, 16 minutes on unhypered Kodak Pro 400 PPF; April 4, 1997)

magnitude) and its lengthy tail, made a big difference in the comet's public impact. Over the rest of March and April, it's likely that more people saw Comet Hale-Bopp than have ever seen any comet in the past, including Hyakutake and Halley. Extensive news coverage brought millions of people outdoors in the evenings to see the wonder in the northwestern sky. And it was easy: anyone who stepped outside and glanced even casually upward could spot the comet and feel its thrill.

Many onlookers contrasted the appearance of Hale-Bopp with that of Hyakutake the year before. Two differences stood out immediately, the

On a trip into the California desert, Rick Scott shot the comet in the spiky branches of one of Joshua Tree National Park's namesake plants. He put his camera on a tripod and used a 40 mm f/2 lens plus a 30-second exposure on Kodak Ektachrome P1,600. (April 4, 1997)

Most people saw Comet Hale-Bopp in the evening sky from late March through mid-April 1997. David Lane photographed it over trees from St. Croix, Nova Scotia, Canada. He used an 8-minute exposure on Kodak Royal Gold 1,000 with a 50 mm f/2.8 lens. Because he tracked on the sky, the trees are blurred. (April 8, 1997)

first being Hale-Bopp's bland color. Hyakutake shone with a startling pale turquoise, thanks to strong emission from diatomic carbon (C_2) in its coma. Hale-Bopp, however, looked plain white or creamy-white. Much of its light was in fact ordinary sunlight reflecting off the fog of dust filling its coma and streaming off to form the dust tail.

The second big difference lay in the nature of the two comets' tails. All comets display a tail of ionized gas and one of dust, but the relative prominence of each depends on the particular comet, its composition, and its activity. For Hyakutake, the bluish gas tail (loaded with carbon monoxide) was by far the more visible and lengthy. On the nights when Hyakutake came closest to Earth, its tail appeared to stretch more than 60° long — across a third of the entire sky. For Hale-Bopp, however, the most prominent tail was the one made of dust. Even when seen under the most favorable conditions, this remained much shorter than Hyakutake's — estimates placed its length at 15° to 20°. (Hale-Bopp's gas tail actually extended a little farther, about 20° to 25°, but because its strong blue color was less detectable to the eye, few observers were impressed by it. Photos however showed it more clearly.)

The physical differences between the two comets led to the curious result that how people rated them depended on where they saw them. Thanks to the skyglow from man-made lighting that hangs over every city, Hyakutake's long gas tail passed unnoticed by most urban viewers. For them Hale-Bopp's bright inner dust tail stood out better. However for observers under dark skies, which included most seasoned comet-watchers, the long gas tail of Hyakutake was far more impressive than the shorter (if brighter) dust tail belonging to Hale-Bopp. Another factor in how both comets fared was the length of time that each was visible at peak brightness. Hyakutake's entire apparition lasted just a few weeks and it spent only a couple of days at its best. Hale-Bopp, on the other hand, was a striking object in the morning or evening sky for about two months. Coupled with its leisurely advent, this gave Hale-Bopp a clear edge in the publicity game and, for that reason if no other, it was seen by many more people.

On March 22, 1997, Comet Hale-Bopp made its closest approach to Earth, passing us at a distance of 1.31 AU (122 million miles; 197 million km). It appeared about –0.7 magnitude in brightness with a dust tail

Activity in the coma continued as Hale-Bopp drew away from the Sun. To bring out streamers and jets, Ian Griffin of the Astronaut Memorial Planetarium and Observatory in Cocoa, Florida, combined 25 electronic photos taken through a Lumicon Swann Band filter, and processed them to reveal inner details. Griffin used a 12-inch Maksutov telescope with an SBIG ST8 CCD working in medium resolution mode. (May 8, 1997)

about 9° long. Over the next few days, Hale-Bopp glided 5° north of M31, the Great Galaxy in Andromeda — although a Full Moon on the 24th kept the sky bright, reducing the spectacle a bit. (At Full Moon, a deep partial lunar eclipse visible mainly throughout the Americas somewhat dimmed the Moon for an hour.) On April 1, the comet passed perihelion at a distance of 0.914 AU (85 million miles; 137 million km). It was traveling at a velocity of 27 miles per second (44 km/s). At around this time, the satellite Solar and Heliospheric Observatory (SOHO) detected a halo of thin hydrogen gas around the comet that was as big in diameter as the distance from the Earth to the Sun. On the ground, observers' estimates of the comet's brightness varied a lot, but averaged around –0.7 magnitude or slightly brighter. The coma appeared about 15 arcminutes across, and observers under dark skies could trace a tail spanning 15° to 20°. Both gas and dust tails were visible, although the gas tail shone a lot fainter.

In the first week of April, the comet moved from Andromeda into Perseus, aiming toward the Pleiades star cluster in Taurus. By an odd coincidence, on April 9, Comet Hale-Bopp passed about half a degree from the place on the sky where Comet Hyakutake had been exactly one year to the day before. Although a full year separated their passages, photos show both comets lying near the bright variable star Algol, Beta (β) Persei. (Of course, the comets passed Earth at greatly different distances.)

After Full Moon on April 22, Comet Hale-Bopp glided above the Pleiades as they sank into the western evening sky. The bright Moon reduced the comet's apparent visibility: the gas tail was all but invisible to the eye (though it appeared more distinct in photos) and the dust tail shrank to about 10° long or less. The comet's estimated brightness also dropped to about zero magnitude. On the 25th, the comet lay 1 AU from the Sun and 1.6 AU from Earth. Its distance was growing, and the best part of the comet's apparition was over. Dropping lower each evening, Hale-Bopp cut across Taurus, moving towards Orion. On May 8, a slim crescent Moon passed south of the comet, making a fine scene for binocular viewers and photographers. At this point, the comet's dust tail was still 5° to 10° (for observers away from city lights), and the comet's overall brightness was about magnitude 0.3. But Hale-Bopp was sinking rapidly in the west. By late May, evening twilight had swallowed both the comet and Orion.

For viewers in the Southern Hemisphere, the apparition's best part began in mid-May, just about the time it was ending for the Northern Hemisphere. In the first few days of June, as Orion stood in the western sky after sunset each evening, the comet was passing Betelgeuse. But its brightness was down to 2.5 magnitude and the dust tail had shrunk to only a degree or two long. With the disappearance of Orion for southern observers, the comet vanished into the sunlight.

Southern observers picked it up again in the morning sky around the middle of July, with the comet being about 3.3 magnitude and having a half-degree long tail. As weeks followed, Comet Hale-Bopp continued its slow drift back into obscurity. Its brightness sank steadily toward magnitude 4 and 5, and the coma and tail likewise diminished. Northern Hemisphere observers caught a final glimpse at the comet in

Contortions in Hale-Bopp's gas tail show clearly in this image taken by Stephen Larson of the Lunar and Planetary Laboratory, University of Arizona. He used a filter that suppressed light from the dust while bringing out emission from ionized carbon monoxide. (S. Larson, LPL/ University of Arizona, Tumamoc Hill Observatory; May 7, 1997)

Farewell Hale-Bopp! Gordon Garradd's electronic image was taken over a year after perihelion, when the comet lay 5.4 astronomical units from the Sun. Activity is steadily dwindling but still present. Garradd's photo, taken with a 10-inch Newtonian reflector and a Hi-Sis 22 CCD camera fromLoomberah, New South Wales, Australia, shows a "fountain" of dust on one side of the nucleus. (May 23, 1998)

late September and October, but the comet was very low in the sky, dim, and hard to see.

Over the following months, the comet moved steadily southward on the sky, passing into Puppis and Carina and finishing up the year south of Canopus. The last naked-eye sighting came in early December 1997. As 1998 began, the comet was visible only in binoculars or telescopes. During January, its brightness faded through 8th magnitude, with a coma only 5 or 6 arc-minutes across and a tail extending about 1.5°. Interestingly, an "anti-tail" showed up in long-exposure photos. This is an optical effect, produced when dust lagging behind in the comet's orbital plane appears to extend from the nucleus toward the Sun.

By late May 1998 the comet was approaching 10th magnitude and dwindling steadily, while showing a small coma about 4 arc-minutes across. But Hale-Bopp is not just quietly disappearing. In August, when the comet was more than 6 AU from the Sun, astronomers at the European Southern Observatory detected strong emission from methanol and hydrogen cyanide, which had never before been found in any comet that far out. In October 1998, when the comet was 6.7 AU from the Sun, observers imaged its coma, which was about 3 arcminutes across and at 10th magnitude. The coma's marked asymmetry showed that jets were still active on the nucleus.

How long will Hale-Bopp remain visible? Amateurs will be able to follow it for several years into the new century, the length of time depending on the equipment they have. Current telescopes and detectors should let professional comet observers track Hale-Bopp fairly easily until about 2020, when the comet lies 43 AU from the Sun and its brightness is about magnitude 29 or 30. But by around 2050, when it lies 75 AU out and has a brightness of 32nd magnitude, the comet will be hard to tell apart from the many distant faint galaxies that crowd the sky.

For planetary scientists, two Great Comets arriving a year apart was the highlight (and the surprise) of a professional lifetime. Among the startling new findings for Hale-Bopp was the discovery of a new kind of tail made of neutral sodium atoms. While hints of sodium had been detected in Comet Mrkos back in 1957, in April of 1997 a multinational team of European observers photographed an unmistakable (if faint) tail made of glowing yellow sodium. The tail ran for at least 17 million miles (28 million km), and was some 440,000 miles (700,000 km) wide. It grew out of the coma and lay between the dust and gas tails, hinting that it had a mass lighter than the dust, but heavier than the gas. Meanwhile, other observers discovered that the dust tail itself contained sodium atoms, and theorists proposed that this was the source for at least some of the sodium that wound up in the sodium tail.

Even the dust revealed new aspects. Observations from the ground and space uncovered specific types of silicates in the dust tail of the comet, including olivines and some pyroxene-rich minerals. These are quite similar to the interplanetary dust particles that enter Earth's atmosphere. At peak activity, Hale-Bopp was shedding over 400 tons of dust per second. This, as noted before, was more than 100 times Halley's peak rate. Yet so large is Hale-Bopp's nucleus — 25 miles (40 km) in diameter — that it probably lost less than 0.1% of its material on this trip around the Sun.

Controversial results suggest that the nucleus may have spawned

several large fragments — "baby Bopps," in effect. The evidence lies in a few Hubble Space Telescope photos that appear to show moving bright spots deep in the dust coma. If these are real — and not, for instance, optical illusions caused by intersecting streams of material from jets — then the fragments would further underscore the comet's high rate of activity. The jury is still out on this and many comet scientists are skeptical. But it's possible that, as the comet's activity continues to drop and the obscuring coma dwindles and thins, long-exposure photos may reveal more than a single point of light in the nucleus.

While Hale-Bopp was extremely dusty, it also produced a great quantity of gas — between 2 and 5 times as much as the dust, depending on its activity. As the comet came sailing in, the rate at which it was producing carbon monoxide began to slow just as its production of water vapor started to get going in earnest; this occurred at about 3.5 AU, roughly in the middle of the asteroid belt. Solar distance (which governed the amount of heating the comet felt) was the controlling factor in the production of the molecules that evaporated from the comet's ices to form the coma. With the comet now retreating into space, all its activity will gradually shut down. The last-active ingredients will probably be carbon monoxide and carbon dioxide, two of the first to become active on the way in.

Observations detected eight molecules never before seen in a comet. These included sulfur monoxide (SO), sulfur dioxide (SO_2), nitrogen sulfide (NS), formic acid (HCOOH), methyl formate (CH_3OCHO), and formamide (NH_2CHO). These joined already-known molecules such as water (H_2O), heavy water (HDO), formaldehyde (H_2CO), hydrogen sulfide (H_2S), carbon monosulfide (CS), carbonyl sulfide (OCS), cyanogen (CN), methyl cyanide (CH_3CN), hydrogen isocyanide (HNC), methanol (CH_3OH), acetylene (C_2H_2), ethane (C_2H_6), ammonia (NH_3), methane (CH_4), and others. Hale-Bopp's brightness means that for the first time astronomers have the data that will let them create maps of the coma showing where the various species were most active and how this changed with the comet's distance from the Sun.

One of the most complicated problems in all comet science today is tracing the various chemical reactions that lead to the substances seen in the coma and tail. Some atoms and molecules are the products of

For most viewers in the Northern Hemisphere, the comet's apparition ended in May 1997. Michel Benvenuto took his last photos of Hale-Bopp on May 8, from Mont Chauve, France, when he caught the comet with a young Moon below it. He used a 210 mm f/5.6 lens and a 5 second exposure on Kodak Ektar 1,000. (Note the lunar seas visible in the earthlight.)

complex molecules being broken up by solar ultraviolet light. Others come from the recombination of chemical fragments. Difficulties mount in studying the problem because of the rich chemical "stew" that comets can contain, and the multitudinous ways the atoms and molecules can react with each other. These reactions vary in turn depending on the comet's distance from the Sun and its associated warmth. Hale-Bopp has provided a lot more information to work with and if this complicates the problem in the short term, it will produce a much more realistic picture in the long run.

The composition of the dust and gas of Hale-Bopp indicated that it shares a common origin with other solar system bodies, including Comet Halley. This was not unexpected, but some comets — Hyakutake, for example — contain ices whose makeup points to an origin in a different interstellar cloud than that which formed the solar system. This doesn't necessarily make Hyakutake an interloper thrown out of some other star's Oort Cloud. But some of its material could well have a non-solar-system origin and have been mixed into the debris out of which Hyakutake as a whole formed. This is not the case for Hale-Bopp: on the evidence of this passage at least, Hale-Bopp appears to be a completely home-grown product.

Comet Hyakutake gave scientists a huge surprise when they discovered X-rays coming from its coma. They then raced to check out Hale-Bopp and other comets — and found X-rays from these objects too. But even with several examples in hand, detection has proven much easier than explanation. After a lot of debate, two theories to explain the X-rays hold the lead for the time being. The first says that X-rays fly from a comet when heavy ionized atoms in the solar wind slam at high speed into the lighter atoms of the coma. The other theory holds that extremely fine dust particles in the coma may reflect X-rays coming from the Sun, like a kind of imperfect mirror. Neither theory can explain everything that has been observed, and both have drawbacks. The final answer is likely to incorporate parts of both theories, and further mechanisms may emerge from longer study and more cometary examples.

One of Hale-Bopp's greatest gifts was its powerful activity, which helped Alan Hale and Tom Bopp discover it almost 20 months before perihelion. This gave comet astronomers, amateur skywatchers, and ordinary people an unprecedently long look at a beautiful Great Comet. Now, as Comet Hale-Bopp heads back into deep space whence it came, scientists are looking to integrate the mountain of data with their various models of how comets work. The next step includes missions to comets that will examine and sample several comets at close range. (See the next chapter.) The end result should be a vastly better understanding of these ancient relics of the solar system's earliest days — and a better appreciation for the next Great Comet that sails out of the night into our view.

5 *Space Missions to Comets*

If there's one lesson the Space Age has taught, it's that we don't really know another solar system body until a spacecraft pays it a visit. After centuries of ground-based studies using telescopes, scientists had learned enough to let them sketch the planets and moons of the Sun's family in outline. But that's as far as it went. Only in the last 35 years, as instrumented probes began close-up reconnaissance, has planetary science broken free of the age-old rut which saw it squeezing smidgens of ambiguous information out of blurry telescopic views. When spaceflight opened up this new frontier, scientists found long-standing debates settled almost overnight. And a new understanding grounded in reality began to displace the host of familiar suppositions and guesses that had endured for decades because no one had enough data to challenge them. For a dramatic example of the importance of spacecraft look no farther than the case of Mars.

In 1964 the image of Mars in the minds of most people, including some scientists, featured a world not all that far removed from something that Percival Lowell (of canals fame) might have recognized since it hadn't changed much in six or seven decades. From Earth, astronomers had seen that Mars had ice caps and tawny deserts of blowing sand. While most of them knew that Lowell's canals were fictitious, many felt that Mars probably harbored primitive vegetation or even somewhat more advanced forms of life.

Less than a decade later, however, this comfortable image of Mars, familiar from countless works of popular science and science fiction, had been completely exploded — *twice*. The first bang came in 1965. A few coarse photos taken by the Mariner 4 spacecraft as it swiftly flew past the planet showed a heavily cratered surface, thus portraying Mars as a dry, barren, Moon-like body. Abruptly, Martian life looked a lot more doubtful. The second bang came a little later. By 1972 Mariner 9, parked in orbit around Mars with much better cameras, was photographing a planet that looked startlingly complex. Mars still had craters,

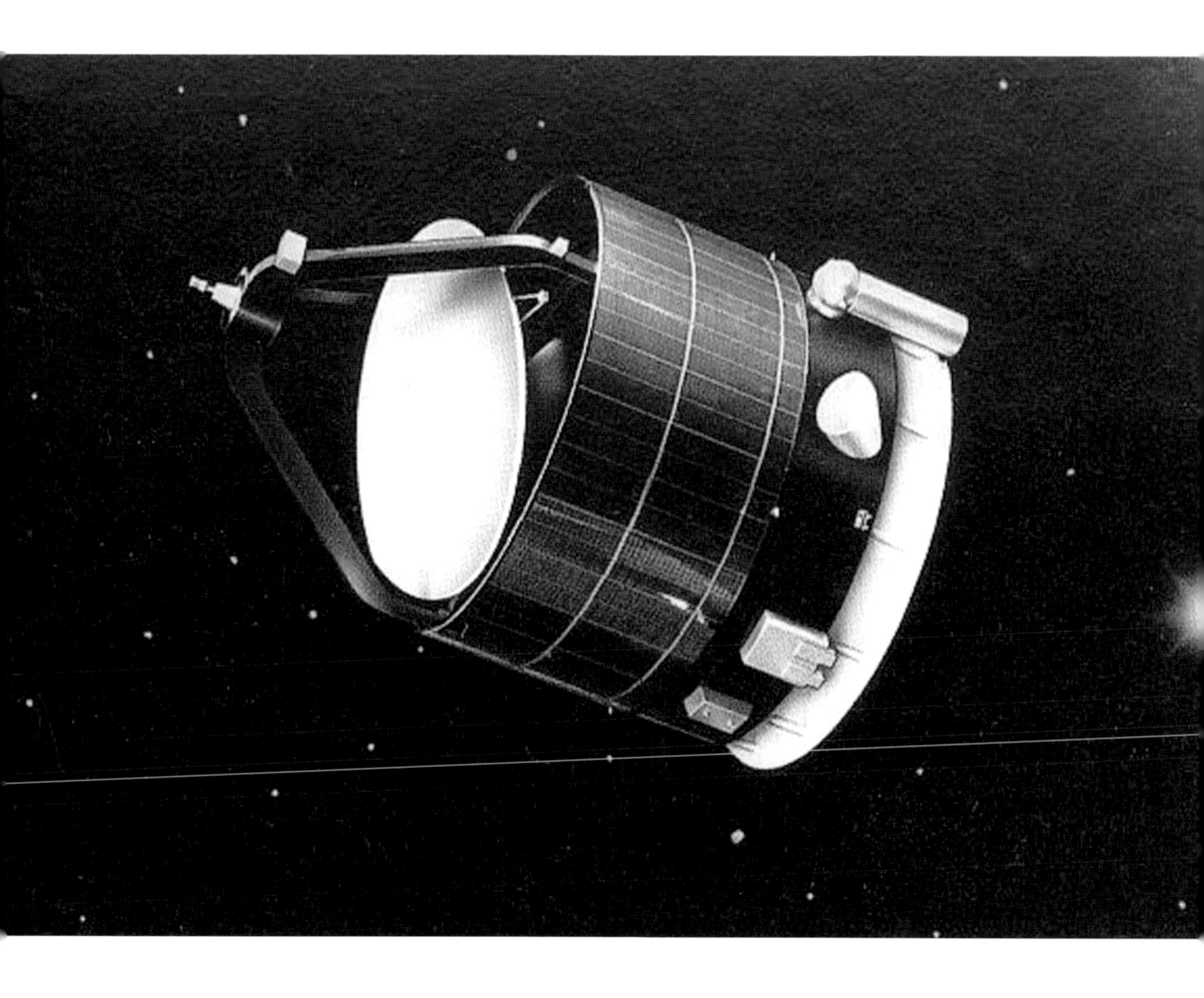

Off to a comet. The return of Halley's Comet in 1986 drew an international fleet of spacecraft to study it at close range. The spacecraft that went closest — and took the most damage — was the European probe named Giotto. (NSSDC, NASA/ Goddard Space Flight Center)

lots of them. But it also had giant volcanoes, yawning canyons with eroded walls, and streamlined islands in vast, long channels down which water unquestionably once flowed. With the geological picture growing more detailed, scientists' ideas about Martian life took another flipflop, to settle roughly where they are today: on the idea that Mars could well have developed life in its earliest period but that present-day life is highly unlikely.

The important point to note here doesn't concern Martian life, how-

Comet Halley, as photographed on March 11, 1986, at the European Southern Observatory. (International Halley Watch/NASA)

ever. It's that there's just no substitute for getting instruments close to a celestial body if you want to know what makes it go.

Unfortunately, comets got into the space probe game fairly late, because of reasons we'll discuss in a little while. For comet science,

Loomings. When Giotto shot past Comet Halley at more than 40 miles per second, it gave scientists their first-ever looks at a comet nucleus. The first image here was taken 6 minutes before closest approach (upper left) and the last at 55 seconds before closest approach (bottom right). (Copyright 1986, Max-Planck-Institut für Aeronomie, Lindau/ H. Uwe Keller)

the Space Age began in 1970 when the **second Orbiting Astronomical Observatory** (OAO-2) observed a huge cloud of hydrogen (H) and hydroxyl (OH) surrounding the head of Comet Tago-Sato-Kosaka. The cloud, created by the breakup of water molecules in the comet's coma, was vastly larger than the comet itself, having a diameter of roughly 10 million miles (15 million km). Similar clouds of hydrogen were spotted around Comet Bennett and Comet Encke that same year by scientists working with the fifth **Orbiting Geophysical Observatory**, OGO-5. The existence of such giant hydrogen clouds around comets had been predicted for several years. But despite their enormous size, the clouds had not been detected earlier because their characteristic ultraviolet wavelengths were blocked by Earth's atmosphere. Only instruments lifted above the obscuring air could see them.

A few years later, when Comet Kohoutek flew past Earth, the comet's 10-month warning time provided an opportunity to mount a coordinated campaign. Besides ground-based observations, scientists examined the comet from space. At the time NASA's **Skylab** space station was in orbit, and the astronaut crew was directed to turn its instruments toward the comet. At a time when the comet was too near the Sun

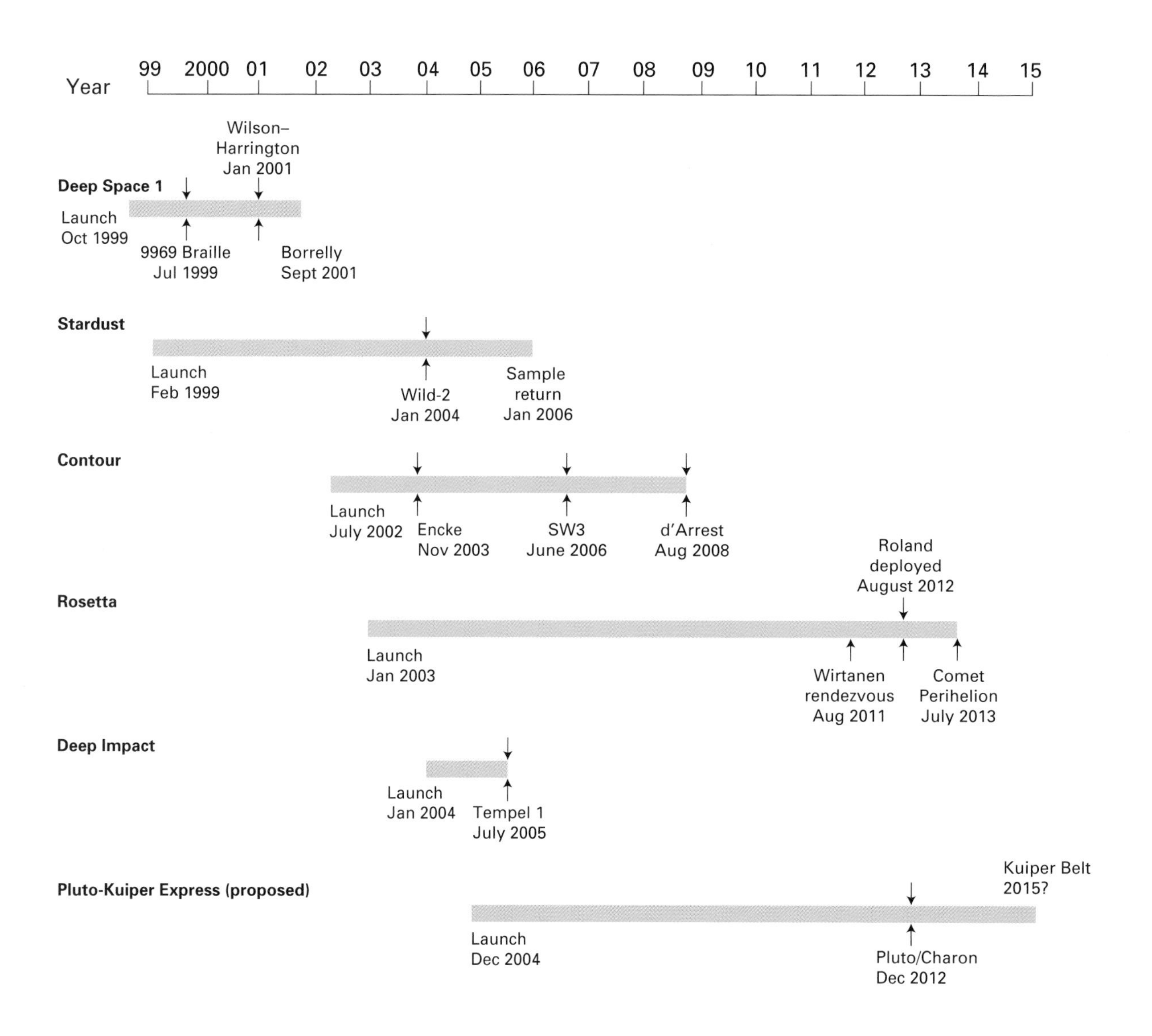

For nearly 15 years following Comet Halley's return in 1986, no missions to comets were launched. But the long spell of inactivity is coming to a close, with new missions already sent off or being readied for flight.

to be seen clearly from the ground, they even made pencil sketches of what they could see by eye and with 10-power binoculars out the station's window and on spacewalks. In addition to the Skylab efforts, sounding rockets were shot up into space to snatch brief looks at Kohoutek before falling back to Earth. The comet was also studied by the **third Orbiting Astronomical Observatory** (OAO-3) and the **Mariner 10** spaceprobe on its flight to Mercury. The result of the campaign was the detection of several new molecules in the comet and a much better understanding of how comet tails (both gas and dust) respond to the influence of sunlight and the solar wind. The upshot is that scientists were far happier about Kohoutek's visit than the general public was!

The new findings proclaimed the value of space-borne observations of comets. In 1978 two new spacecraft, both with ultraviolet telescopes and detectors, were launched and, while neither was specifically designed to study comets, they both made contributions. The first was the International Ultraviolet Explorer (IUE), an astrophysical observatory put in Earth orbit to survey the sky at ultraviolet wavelengths. The other craft, the Pioneer Venus Orbiter, was sent to study that planet's atmosphere from above and map its unseen topography using the spacecraft's radar altimeter.

The **International Ultraviolet Explorer** had a remarkably long lifetime: launched in January 1978, it was finally switched off in September 1996 when the cost of keeping it going would start to pinch newer science missions. IUE created a revolution in astrophysics as it gave astronomers their best-yet looks at some of the more energetic events that occur in stars and galaxies. Comet science didn't rank high on IUE's agenda, but the spacecraft made important observations all the same. During its lifetime, IUE studied more than two dozen comets, most of them faint but not all. In fact, IUE's last comet research included observations of Comet Hyakutake during March 1996. Scientists using IUE tracked the nucleus of Hyakutake for five days, making exposures of up to five hours long. The comet was found to be ejecting 10 tons of water every second as it passed near the Sun. These findings and others provided new insights into the chemical processes taking place inside the comet. IUE observers also confirmed that an apparent "breakup" of the comet's nucleus on March 25, 1996, involved only a small piece of the comet. With other comets, IUE discovered the existence of certain kinds of sulfur and it drew newly detailed portraits of how a comet's coma reacts to the heat and energy from the Sun.

The contributions of the **Pioneer Venus Orbiter** were smaller but no less valued. Like IUE, it also enjoyed a long life (until October 1992), although nearly all of it was spent on its prime task, studying Venus. But, if the craft's comet-observation opportunities were few, its ultraviolet camera still mapped the gas emission from several comets.

As the 1980s opened, NASA was mulling over several dedicated missions to fly a spacecraft close to a comet — and no one needed reminding that Comet Halley was approaching swiftly, as it headed for perihelion in March 1986. Among the mission ideas discussed was a plan to rendezvous with Halley about 50 days before perihelion and travel with it through perihelion and back out to 3 or 4 astronomical units. Tight budgets, however, stopped this idea. Another mission scenario called for using a new ion propulsion drive to send a spacecraft past Halley (and drop off a surface lander) and continue on to a rendezvous with Comet Tempel 2. This ambitious mission fell victim to money woes too — and also a shortening timeline in which to develop it. An alternative mission plan then emerged for a simple, high-speed flight past Halley using existing instruments and spacecraft systems. Frustratingly, this too had to be abandoned when budget cutbacks forced NASA to choose between this (or any) Halley mission and a Venus-orbiting radar mapper. After much agonizing, NASA picked the latter and, following budget and scheduling problems of its own, the Venus mapping mission eventually flew as Magellan. Weighing heavily in NASA's decision to forego a Halley probe was the fact that an international "armada" of five other spacecraft (see below) was in development to greet the comet on its arrival.

NASA did, however, achieve a comet flyby mission and flew it in September 1985, shortly before Halley appeared on the scene! The target was Comet Giacobini-Zinner and the spacecraft was the **International Comet Explorer** (ICE). This probe was originally named International Sun-Earth Explorer 3, and it was one of three satellites launched several years earlier to monitor the interaction of the solar wind with Earth's magnetosphere. Getting the spacecraft from its existing orbit to one that could pass close to the comet took multiple flybys of Earth's Moon, each adding to the spacecraft's velocity. Finally, after one last, low pass over the lunar surface, the newly renamed spacecraft headed off for its rendezvous.

Since ICE's instruments were designed to explore the solar wind and the terrestrial magnetosphere, the spacecraft carried no cameras, just instruments to detect and measure plasma, particles, and magnetic fields. Given its specialized payload, scientists directed the spacecraft to fly though the comet's gas tail about 4,800 miles (7,800 km) downstream from the nucleus. Shooting past the comet at about 13 miles per

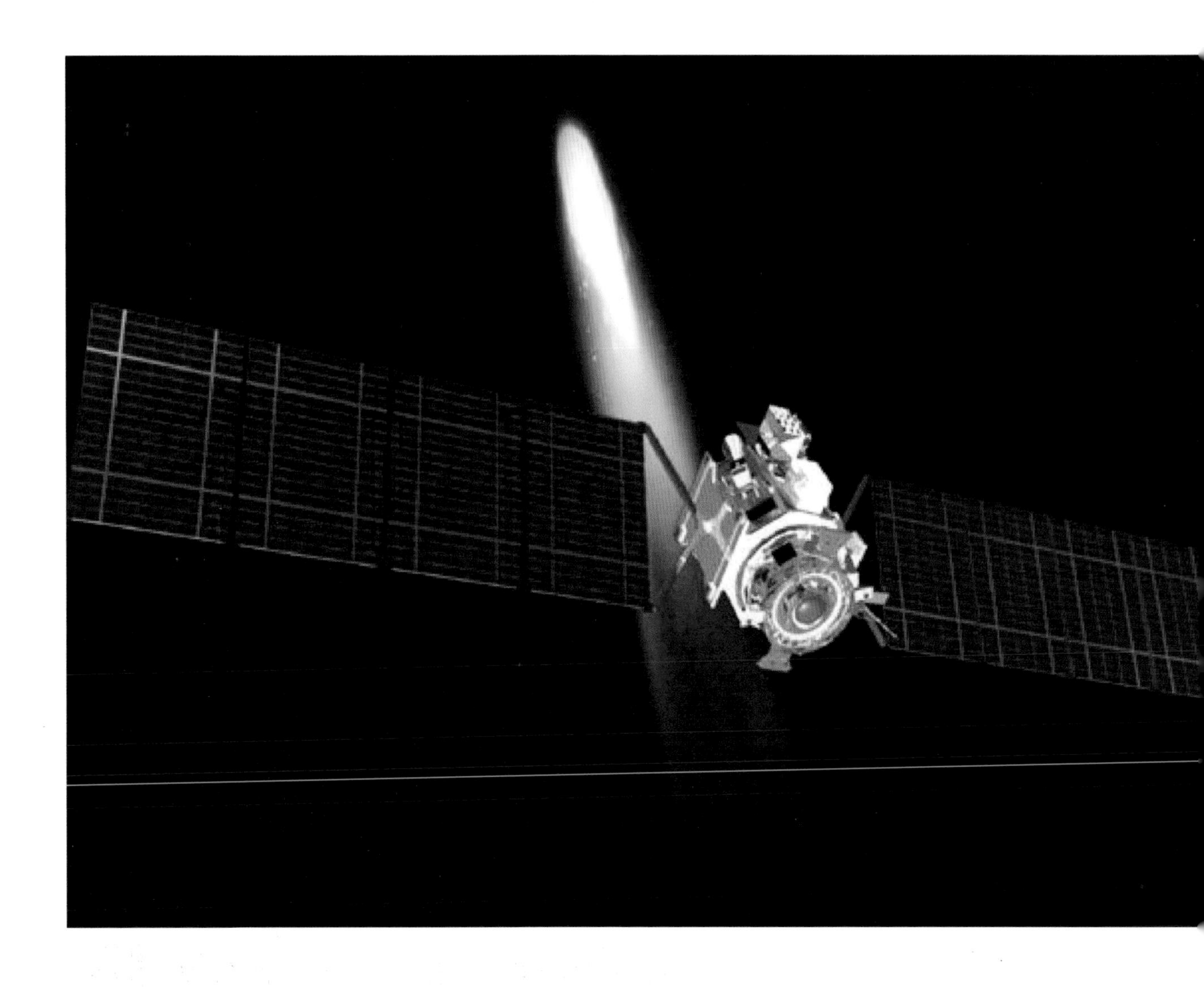

This is only a test... The main reason behind NASA's Deep Space 1 mission is to check out a new propulsion system and instruments. But the mission, launched in October 1998, visited an asteroid in 1999 and, if an extended mission is funded, it will fly past Comets Wilson-Harrington and Borrelly in 2001. (NASA/Jet Propulsion Laboratory)

second (21 km/s), ICE survived its passage with no detectable damage. It identified water and carbon monoxide ions, and found the solar wind's magnetic field wrapped snugly around the comet's nucleus like a hairpin with its open end pointing away from the Sun. The field had a strength only 1/500th that of Earth's field, however. After its encounter with Comet Giacobini-Zinner, ICE flew on to monitor the solar wind upstream of Comet Halley at a distance of about 20 million miles on its sunward side. This gave scientists working with the main Halley fleet advance warning of solar wind conditions that would soon impinge on the comet.

Circling the Sun once every 6.9 years, Comet Borrelly — a possible target for Deep Space 1 — was discovered in 1904 and travels between the orbits of Mars and Jupiter. The image shows artifacts caused by the electronic sensor used to photograph the comet. (James Scotti, Spacewatch Project, University of Arizona)

(Nor has ICE's story ended yet. It will return close to the Moon once more on August 10, 2014, at which time NASA may send a mission to recover it. This would let them examine its solar panels and structures for signs of damage by interplanetary and cometary dust particles. And should ICE be retrieved, it won't be junked after the dust studies are complete — NASA has already donated the spacecraft to the Smithsonian Institution.)

NASA's International Comet Explorer was largely a preamble to the main Halley show. The multinational Halley fleet contained five spacecraft — two Japanese, two Soviet, and one European. Unlike ICE at Giacobini-Zinner, all these craft passed on the sunward side of Halley's nucleus, although their encounter distances varied greatly. The two that passed most distant were **Suisei** and **Sakigake**, probes launched by

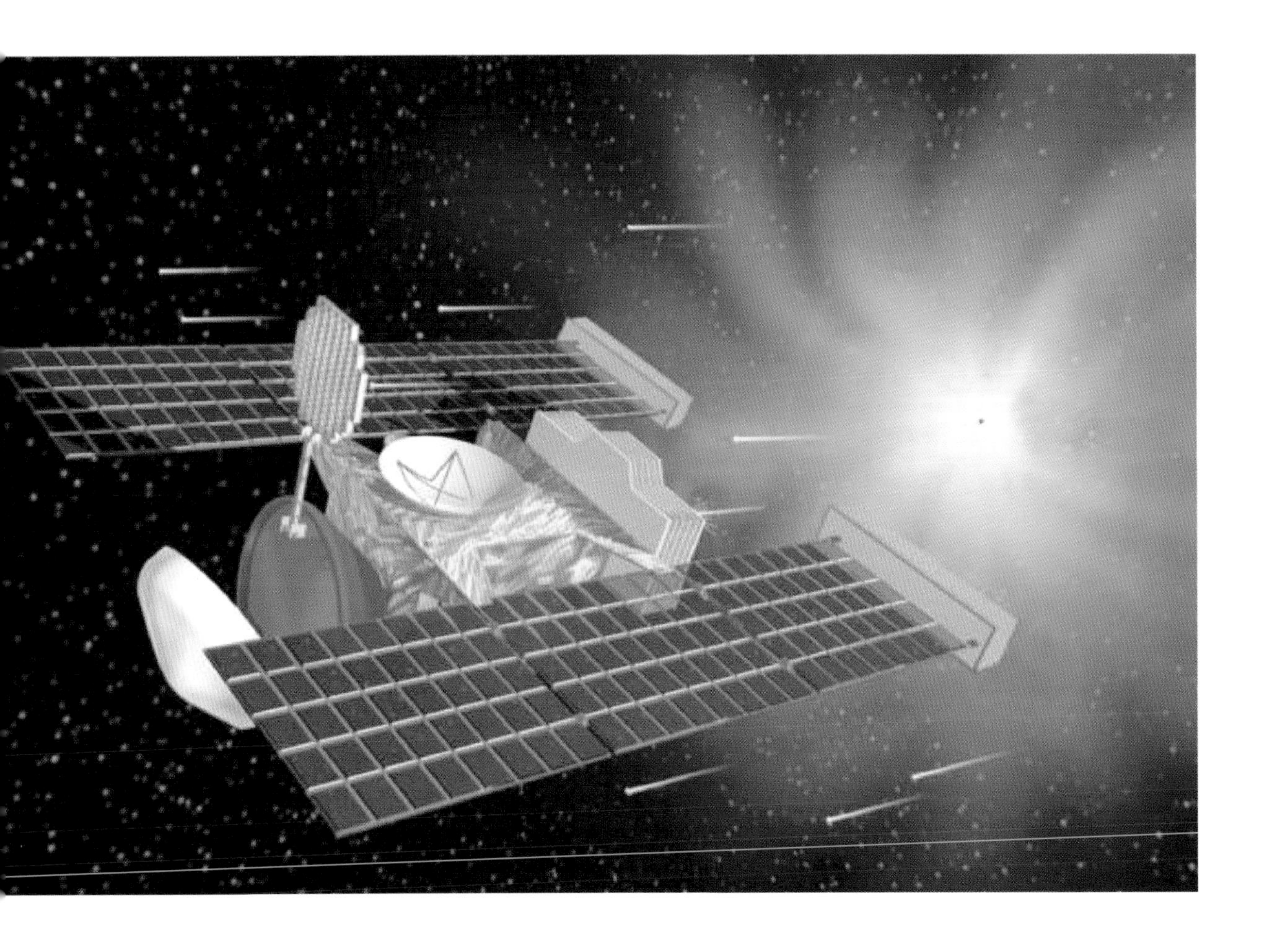

...And put it in your pocket. NASA launched a mission named Stardust in February 1999 to go to Comet Wild 2. When it makes rendezvous with the comet in 2004, the spacecraft will fly within 60 miles (100 km) of the nucleus. Using the honeycomb paddle seen at left, it will collect microscopic samples of the comet's gas and dust tails, bringing them back to Earth for analysis. (NASA/Jet Propulsion Laboratory)

Japan. (Suisei means "comet" in Japanese, and Sakigake means "pioneer.") The two spacecraft differed in their instrument packages, with Suisei carrying an ultraviolet camera to observe the hydrogen cloud and to hunt for specific ions in the coma and tail. Sakigake's mission focused on measuring the solar wind near the comet, detecting radio emission from molecules and particle-field interactions, and surveying ions in the solar wind. Suisei passed 94,000 miles (151,000 km) away from the nucleus on March 8, 1986, while Sakigake's closest approach came on March 11, at a distance of 4 million miles (7 million km).

Suisei detected cometary ions of water, carbon monoxide, and carbon dioxide. It also struck two dust particles and survived. Sakigake found the solar wind was affected by the comet, even at the large distance at

Comet Wild 2. Stardust's target comet is a recent arrival from the colder parts of the solar system, and now moves between the orbits of Mars and Jupiter. Scientists expect to see evidence of new activity on it when the spacecraft makes rendezvous. (NASA/Jet Propulsion Laboratory)

which the probe passed. (After the encounter, Suisei and Sakigake remained active but have run out of maneuvering fuel and both are now considered dead by mission controllers.)

The two Soviet probes, **Vega 1** and **Vega 2**, arrived at Halley on March 6 and March 9 respectively, passing 5,500 miles (8,900 km) and 5,000 miles (8,000 km) from the nucleus. Their instrument packages photographed the nucleus and studied the spectroscopic composition of the nucleus and coma. They also carried detectors to measure the dust's composition and particle sizes. (Interestingly, the flight of the Vega craft to Halley took them past Venus on the way to the comet. So Soviet mission planners prepared separate probes for Venus as part of the total payload. These were dropped off as each Vega spacecraft swung past the planet. Both Venus instrument packages worked successfully.)

Vega 1's photos located the nucleus of Halley inside the gassy coma. This helped refine the trajectory that the European probe Giotto followed a week later on its much closer passage. Vega 1 also found what turned out to be two jets on the nucleus, and measured the surface to be at room temperature or higher, indicating a dark coating of Sun-warmed dust. The dust itself had a carbonaceous composition. A couple

The Comet Nucleus Tour — the Contour mission — is due for launch in July 2002 and features flexibility. Encounters with three short-period comets are scheduled but after the first flyby (Comet Encke), mission controllers can change the flight plan to send the spacecraft to other known comets — or retarget it to meet a newly discovered one that comes within range. (Applied Physics Laboratory, Johns Hopkins University; Cornell University)

of days later, Vega 2 made a closer passage and photographed the nucleus in greater detail.

But of the whole fleet of spacecraft perhaps the most spectacular results came from the European Space Agency's **Giotto** mission. This multinational spacecraft made its closest approach to Comet Halley on March 13, 1986. The name of the spacecraft honored the famed artist Giotto di Bondone, who painted the star of Bethlehem as a comet, possibly Halley's at its apparition in 1301.

The spacecraft's flight plan was fast, close, and risky. Giotto was to go as near to the nucleus as possible without going so close as to be certainly destroyed. But because no one knew how much dust filled the coma — after all, finding this out was one of the mission goals — there was no guarantee that Giotto would survive, despite its sturdy dust shield. Given the relative speeds of 42 miles per second (68 km/s) between the probe and the comet, hitting even a sandgrain-size piece of dust could severely damage the spacecraft. In the end, Giotto flew past the nucleus on its sunward side at a distance of 370 miles (596 km). It photographed details as small as 100 yards across on the dark oblong nucleus that measured about 10 miles by 5.2 miles by 5.1 miles (16 km

The second known periodic comet, Comet Encke, is an older, well-evolved body with a nucleus about 2 or 3 miles in diameter. It is the first and primary target for the Contour spacecraft, which is scheduled to fly past the comet in November 2003. (NASA/Jet Propulsion Laboratory)

by 8.4 km by 8.2 km). Giotto saw two large jets and several small ones venting gases that flowed out to become the coma and tail. (The probe's photos also confirmed that Whipple's dirty-snowball model for a comet nucleus was correct.)

Besides its camera, Giotto carried mass spectrometers, dust analysers, a magnetometer, and plasma instruments. These worked almost perfectly. Dust impacts stayed below predictions until shortly before closest approach, when the spacecraft apparently crossed through a jet of debris. Immediately before it flew closest, Giotto struck a particle big enough — probably about a millimeter in diameter — to set it spinning off-kilter. This reduced the data transmitted, although the spacecraft's thrusters soon corrected the gyrations. Dust impacts also knocked the camera's periscope mirror askew (putting the camera out of action) and shot the star-mapper's baffle full of holes.

Despite the damage it received, the Giotto mission was a tremendous success. (In fact, after its Halley passage, Giotto remained in good enough working condition to be retargeted to another comet, Grigg-Skjellerup. It flew past this faint comet in July 1992.)

At Halley, Giotto had photographed a comet nucleus in detail for the first time, finding seven jets on the surface and measuring their composition. It drew maps of the coma that are still being studied. It told how the comet responds to the solar environment around it, and it painted a far richer picture than before of the hordes of chemical species that blossom and die through complex reactions in the coma and tail. It's no exaggeration to say that, after the Halley encounters, comet science became a whole new ball game.

The Halley flights marked a high point, yet they also stand out in contrast to what followed. In 13-plus years since those action-filled days of March 1986, not a single comet-dedicated mission was launched. It's worth asking why.

Several obstacles stand in the way of sending space missions to comets. Some difficulties originate in nature and others in the Earth-bound realities of spaceflight funding. One hurdle is natural: bright comets appear unpredictably and commonly have short apparitions. When you don't know what's coming at you or from what direction, it's hard to prepare a mission to go out to meet it. No nation has the resources to build a generalized comet probe and put it in mothballs while waiting for a new comet to show up. Comet Hale-Bopp was unusual in giving almost two years' warning of its arrival, and Comet Halley is also an exception because it is bright and has a well-determined orbit. A more typical example is Comet Hyakutake, which made its closest approach to Earth less than two months after discovery. A lead time like that is too short to let scientists prepare any spacecraft for launch.

Second, bright comets typically approach at high velocity on orbital paths tipped at steep angles to the ecliptic — witness again Hyakutake and Hale-Bopp. This technicality makes the mission planner's job much more difficult because a spacecraft has to carry lots of fuel to adjust its trajectory from that of Earth to match that of the comet. This cuts into the instrument payload and boosts the cost of the launch rocket. The only alternative approach, given a short lead time, is to forego any

attempt at a low-speed rendezvous and settle for a high-velocity flyby like the Halley encounters. But as Giotto showed, these pose risks, both to the spacecraft's health and to the amount of science the probe's instruments can snatch from the few minutes or hours around closest approach. Having done high-speed flybys with Halley, comet scientists would now prefer longer encounters to build on what's been learned in the fleeting snapshots grabbed previously.

A third barrier is fiscal reality. With space exploration budgets under severe constraint everywhere, it has proven politically impossible to assemble the money to design, build, and launch a spacecraft essentially "on demand" in time to meet the rare and unpredictable bright comets, given the scant warning that mission planners get. As noted before in Chapter 4, even the 20 months that elapsed between the discovery of Comet Hale-Bopp and its closest approach to Earth were too short to do more than organize a campaign using ground-based telescopes and ones already in flight. The realities of the funding and engineering cycles force comet scientists to send spacecraft to well-known, if less spectacular, comets whose movements can be predicted many years in advance.

In partial acknowledgement of this problem, several years ago NASA initiated a program called Discovery. It is designed to get unmanned solar system missions flying more often and at lower cost. Discovery's basic ground rules are simple: a mission must have outstanding scientific merit, it must take no longer than 35 months from approval to launch, and it must cost no more than $190 million in 1999 dollars. That's a spartan regimen, considering what space missions have cost in the past. Yet it is technically feasible and appears politically supportable — key elements given the stringent fiscal environment NASA must work within these days.

Rosetta Stone for comets? The European Space Agency is planning an ambitious mission named Rosetta to visit Comet Wirtanen. After a launch in January 2003, the spacecraft spends eight years passing Mars and two asteroids before making rendezvous with the comet in 2011. After studying the comet for a year, Rosetta will place a lander (RoLand) on its surface. (European Space Agency)

There are no guaranteed berths in the Discovery program, however. Any proposed comet mission must contend on grounds of merit and technical and fiscal details with all other Discovery proposals submitted during the same funding round. These may include exciting missions to Venus, the asteroids, the Moon, or Mercury, to name just a few. With the small funds available, each year only one or two Discovery-class missions (out of 30 to 40 proposed) will get the green light. And here again, missions to bright new comets remain at a disadvantage because even the accelerated Discovery schedule still runs far longer than most comet apparitions.

But the picture is far from gloomy. Despite the sobering realities, the long post-Halley drought of comet flights is about to end. At present, four missions (plus a potential fifth and sixth) are in development and due for launch by the early 2000s.

The first of the new missions is NASA's **Deep Space 1**, launched in October 1998. Managed by the Jet Propulsion Laboratory for NASA, DS1 is designed mainly to try out a new ion propulsion engine for space probes. Thinking creatively, however, NASA chose to piggyback a science mission on what began as an engineering test. Ion propulsion represents an important new step in spacecraft development, especially for missions that intend to cruise past several targets in succession; this will be the first time it has been used as the main propulsion system in deep space. Also being tested are a solar concentrator power array, an integrated camera and imaging spectrometer, and autonomous optical navigation. The last is the ability of a spacecraft to take navigation photos and plot a course by itself, acting as its own mission control. Such an ability is increasingly essential for spacecraft heading into deep space, where communication with Earth (even at the speed of light) can take hours.

The primary science goal of the DS1 mission was to visit the asteroid 9969 Braille in late July 1999. The asteroid was chosen from more than 100 flyby possibilities. Its elliptical orbit curves within and outside of Mars' orbit of the Sun, at its farthest going out more than 3 astronomical units, roughly to the middle of the main asteroid belt. Although scientists believe the asteroid's diameter is approximately 2 miles (3 km), they know little else about the body. With the flyby, they can learn more about its shape, size, surface composition, mineralogy, terrain, and rotation speed.

By September 1999, Deep Space 1 will have completed its main goal of testing new technologies and performing the asteroid flyby. If DS1 receives funding for an extended mission — which is by no means certain — then it will be put on course for flybys of Comet Wilson-

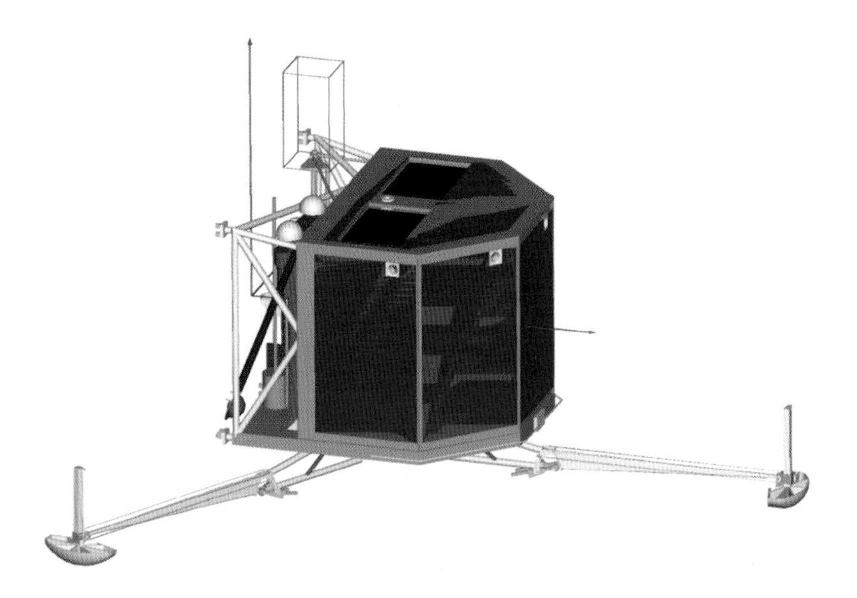

Boldly going. Rosetta's lander, named RoLand, is a boxy craft designed to make on-site measurements of Comet Wirtanen. It will be dropped onto the comet a year before perihelion, and will collect data as the comet approaches the Sun and its activity grows. If the Deep Space 4 mission is cancelled or delayed, RoLand will be the first spacecraft to land on a comet. (European Space Agency)

Harrington in January 2001 and Comet Borrelly in September 2001. Wilson-Harrington is a curious object that appears to be transitional, being part-comet and part-asteroid (see page 75). Comet Borrelly is one of the most active comets that regularly visit the inner solar system.

If it makes rendezvous with these comets, Deep Space 1 will photograph their nuclei, measure their sizes, reflectivities, roughnesses, and activities, and explore the composition of the dust and gas emissions from their nuclei. The spacecraft will also study the composition of the coma and the gas and dust tails of the comets. How close the spacecraft can go to them will depend entirely on how active they are at the time. The more active the comets are, the farther away Deep Space 1 will have to remain, since NASA doesn't intend to finish the flight with the spacecraft hitting any cometary debris.

Gathering debris, however, is exactly the aim of the next comet flight to leave Earth after the launch of Deep Space 1. This is a NASA spacecraft named **Stardust**, a name chosen to reflect the prime goal of the mission, which is to collect samples of dust from Comet Wild 2. The craft will also pick up any interstellar particles that happen to strike its collector. Launched in February 1999, Stardust is a collaboration of JPL, NASA, and the University of Washington. Its flight plan calls for it to cruise

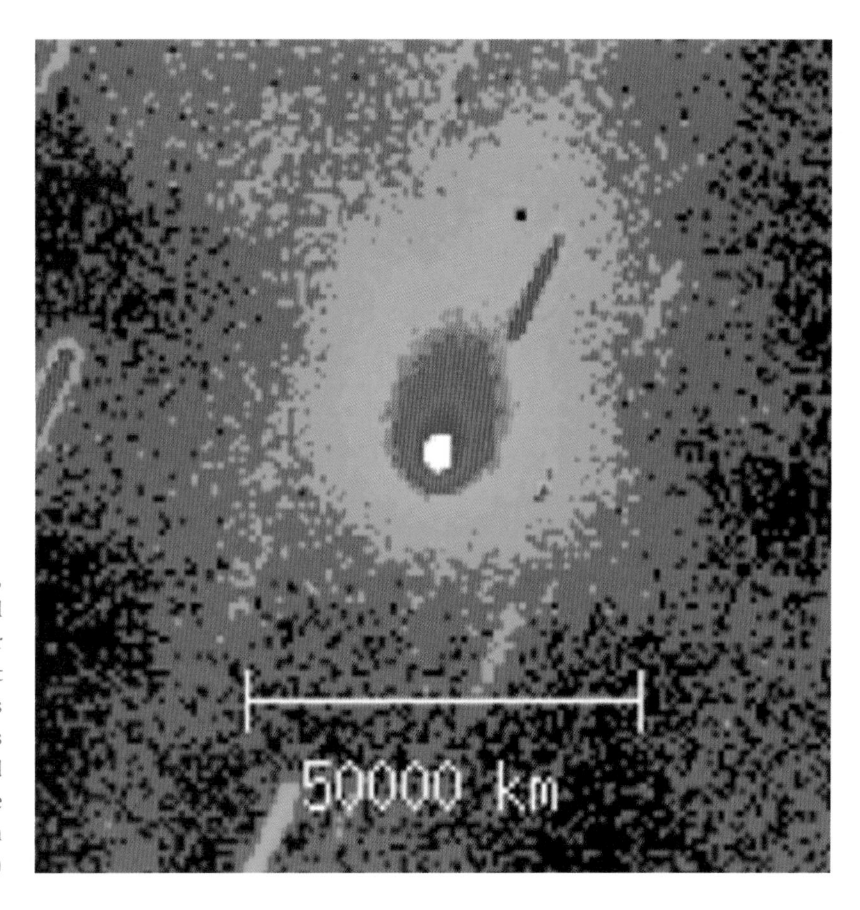

Being relatively quiescent, Comet Wirtanen makes an ideal candidate for placing a lander on its surface, an activity that poses risks to the spacecraft's health. Because the comet's gravity is negligible, RoLand will have to anchor itself to the comet. (ESO/ESA Wirtanen Observing Team)

three times around the Sun before meeting with the comet in January 2004, some 98 days after the comet's perihelion. As it passes about 90 miles (150 km) sunward of Wild 2's nucleus, the camera on Stardust will photograph the nucleus in detail. Since the spacecraft will fly about four times closer to the nucleus than Giotto did with Halley, its camera will reveal surface features that couldn't be detected on Halley. Also, Stardust will view the nucleus over a much wider range of lighting conditions. Plans call for seeing details as small as 100 feet (30 meters) across and, by photographing the scene through special spectral filters, scientists will map areas on the nucleus according to their composition.

But this is only the sideshow. The spacecraft's main mission is to collect dust samples from Wild 2's coma. To do this, the spacecraft will expose a panel coated with a very low density material called aerogel.

This substance is about 60 times denser than air and has been nicknamed "solid smoke." You might liken it to extremely squishy styrofoam or cotton candy. At the relatively slow encounter speeds of 4.8 miles per second (6.1 km/s), aerogel is ideally suited to capture individual particles of comet dust. Because this is the first time any spacecraft has attempted this feat, Stardust will also carry a mass spectrometer to make on-site chemical analyses of the particles that it detects — just in case problems develop with the particle collector.

After the encounter, Stardust will close a hatch to cover the aerogel, thus preserving the specimens intact and safely stowing them in a separate re-entry capsule designed to return them unaltered. Stardust will then slowly orbit back toward Earth, dropping off the capsule as it flies past in January 2006. Plans are for the capsule to come down by parachute over a deserted patch of Utah, some 100 miles (160 km) southwest of Salt Lake City. Recovery teams will pick up the capsule on the ground and whisk the precious comet samples away to NASA's Johnson Space Center in Houston, where they will be studied and analyzed for years to come.

Comet Wild 2 is a fresh import from the colder parts of the solar system. Before encountering Jupiter in September 1974, Wild 2 circled the Sun between Jupiter's orbit and that of Uranus. It now spends its time shuttling between the orbits of Jupiter and Mars, where it is feeling much stronger sunlight and its activity should be correspondingly greater. The Stardust science team are eagerly looking forward to collecting a few pieces of this exotic object and hope that studies of the dust samples will tell them more about how comets and icy planetesimals formed in the solar system's earliest days.

In July 2002, after Stardust is launched but before it returns with its cargo of samples, the **Comet Nucleus Tour — Contour** for short — will depart Earth and make the first of its three planned comet encounters. (See the timeline on page 140.) Contour is not a sample-collection mission like Stardust. Instead, it's designed to explore a variety of comets using cameras and spectrometers. By visiting several comets and seeking their points of similarity and difference, mission scientists hope to learn much more than they would from any one of them.

Contour is a simple and robust probe derived from the successful

Near-Earth Asteroid Rendezvous (NEAR) spacecraft. Designed by Cornell and Johns Hopkins Universities (the team that built NEAR), Contour will carry a small payload of instruments. These are designed to make images of comet nuclei in many colors (wavelengths) and to study the shape and rotation of the nuclei. Contour will also map the location of any jets. The spacecraft will analyse the dust shed by the comets, determining the mass, size, and composition of particles using a mass spectrometer.

All three encounters for Contour are moderately slow flybys. It meets with Comet Encke in November 2003, Comet Schwassmann-Wachmann 3 in June 2006, and Comet d'Arrest in August 2008. The encounters take place relatively close to Earth, so the spacecraft can send back a generous stream of data. Contour's flight plan is also flexible. For example, instead of meeting with Comet d'Arrest for the third encounter, it would be possible to substitute either Comet Wirtanen (in February 2008) or the unusual comet-asteroid Wilson-Harrington (September 2009). If the mission goes to d'Arrest as planned, then there's a chance of a second flyby of Encke. (And if Contour visits Comet Wirtanen, it would survey this comet one apparition before the Rosetta mission — see below — reaches it.) Contour's team also points out that should a newly discovered comet visit the inner solar system while the mission is running, there's a good chance that Contour could reach it. They note that if Contour had been in flight in 1995 and 1996, it could have encountered Comet Hale-Bopp in May 1997.

Flyby speeds won't be quite as low as Stardust's, but much slower than Giotto's frantic dash. They range from 18 miles per second (28 km/s) at Encke to 7 miles per second (12 km/s) at d'Arrest. On closest approach to each comet nucleus, Contour's camera will be able to resolve details as fine as 13 feet (4 meters) across, which is about 25 times better than Giotto at Halley and substantially better than Stardust as well.

This Deep Impact is for real — it's a mission to fire a half-ton projectile into a comet nucleus at high speed and see what happens. When the Deep Impact probe sends its impactor into Comet Tempel 1 in July 2005, scientists expect the crater it makes will reveal what the inside of a comet nucleus looks like. Despite the Hollywood echo, the mission's name was picked long before the movie appeared. (Bull Aerospace)

Each of the three comets was chosen because they are among the largest and most active of the group of short-period comets. Despite similarities, though, they are individually diverse. Encke is an old, evolved comet known to have a low production rate of dust, despite a surface believed to have deep vents, while Schwassmann-Wachmann 3 and d'Arrest have fresher ices and are dustier. Moreover, the nucleus of

Schwassmann-Wachmann 3 split in 1995, which exposed very active ices. It offers a peek deep into the interior of a comet nucleus.

Parallel with the spacecraft activity, the Contour mission will undertake an extensive outreach effort with the public and educators. Project scientists plan to organize a Contour Comet Watch to coordinate the observations of amateur and professional astronomers. This will be modeled after the highly successful International Halley Watch of the 1980s. In addition, workshops for science teachers and students will help them participate in aspects of cometary research and even spacecraft operations during parts of the flight. Lastly, there will be regular updates and releases of news and photos in the mass media and via Internet websites and e-mail. (Such outreach efforts are now routinely required by NASA for all missions where it contributes funding.)

Contour is an impressive and exciting mission. Yet the most ambitious comet project in active development is the European Space Agency's **Rosetta**. This mission is named for the famed black stone found in 1799 near Alexandria whose inscriptions provided the key for understanding Egyptian hieroglyphics. Rosetta's designers expect the mission will do as much for comet studies as the stone did for historians. Rosetta will be launched fairly soon (in January 2003), but it spends eight years getting to its target, Comet Wirtanen. However, the eight-year flight won't pass with the spacecraft just snoozing in hibernation. There are interesting way-stations — in May 2005 Rosetta flies past Mars, in July 2006 it encounters asteroid 4979 Otawara, and in July 2008 it passes asteroid 140 Siwa. Between the Mars and asteroid encounters are two flybys of Earth, in October 2005 and October 2007, which will boost the craft's velocity. (It should be noted that as mission plans are fine-tuned before launch and after, the specific asteroid targets of opportunity may change, but the chosen comet will not. To stay on top of changes, keep in touch with the Rosetta web site listed in Chapter 8.)

The science planned for the cruise phase of the Rosetta mission is relatively simple and low in priority compared with the comet rendezvous and landing. Still, when it flies by Mars the craft will perform remote sensing of the planet, and at the two asteroids, it will pass close enough to image the surfaces in many spectral colors with the intent of mapping mineral compositions. (Since asteroids have proven to be a highly

Orbiting the Sun every 6.5 years, Comet Tempel 1 will be visited by NASA's Deep Impact mission. (NASA/Jet Propulsion Laboratory)

diverse collection of bodies, planetary scientists are trying to make sure that every spacecraft cruising through the asteroid belt takes a look at as many targets of opportunity as the mission can spare.)

The Rosetta mission uses a two-part spacecraft. The larger part, simply called Rosetta, carries instruments for photographing the comet and studying its surface composition. Rosetta also has gas and dust analyzers and plasma detectors. On board is a sophisticated computer intended to make the mission relatively autonomous of mission control on Earth. This is necessary because communications with Earth will be difficult at times during the flight when the spacecraft must carry out some activities.

The second part of the spacecraft is a lander called RoLand (short for Rosetta Lander). This carries a mass spectrometer to analyze the surface materials it touches, along with a camera and various other instruments to detect and identify the gas and dust found on the comet's surface. It also contains an instrument that will try to listen for seismic rumblings inside the comet nucleus.

In November 2011, Rosetta and RoLand make their long-anticipated rendezvous with Comet Wirtanen, joining the comet about 4 AU from the Sun. Then as the comet glides in toward perihelion, warms, and grows active, Rosetta coasts along with it, studying the changes and radioing the data back to Earth. About a year after meeting with Wirtanen, Rosetta will approach the comet closely, go into orbit around it, and deposit RoLand on the surface of the nucleus, believed to be only about 1,300 yards (1.2 km) across and rotating with a 3 to 6 hour period. At that point, the comet will be about 3 AU from the Sun. Then over the following months, Rosetta and RoLand will accompany the comet as it heads for perihelion in July 2013 at a distance from the Sun of 1.06 AU.

The Rosetta mission's chief scientific tasks are to map the comet nucleus in unprecedented detail, analyze its dust and gases, and using RoLand, directly measure the properties of the nucleus at the landing site. The comet's weak gravity means that Rosetta's orbital speed around the nucleus will be less than 1 mile per hour when the probe is 5 miles from the nucleus, and even closer approaches will be easy using the craft's thrusters. Such a slow pace assures project scientists that they will be able to study the nucleus thoroughly and carefully. Nevertheless, there are important unknowns — such as how much dusty debris surrounds Wirtanen and how large the pieces are — which make this a high-risk mission.

Rosetta isn't the last comet flight in the pipeline. One that could surpass all the others for sheer drama is being developed by the University of Maryland and the Jet Propulsion Laboratory under NASA's Discovery program. Named **Deep Impact**, the spacecraft has two parts, one being a 500-kilogram (1,100 pound) copper projectile. The other part is an instrument platform outfitted with both medium- and high-resolution cameras and an infrared spectrometer to study the ejected ice and dust. Deep Impact's flight plan is to fire the projectile into the nucleus of Comet Tempel 1 at a speed of about 6 miles (10 km) per second. Scientists anticipate the resulting explosion will blow a crater in the nucleus at least 80 feet (25 meters) deep and nearly 400 feet (120 meters) across. (Ironically, the mission's name was chosen before the movie *Deep Impact* appeared — but, as mission scientists point out, the name describes exactly what the spacecraft will do.)

Comet Tempel 1, discovered in 1867, is believed to have a nucleus some 4 miles (6 km) in diameter or less, with perhaps 3 percent of

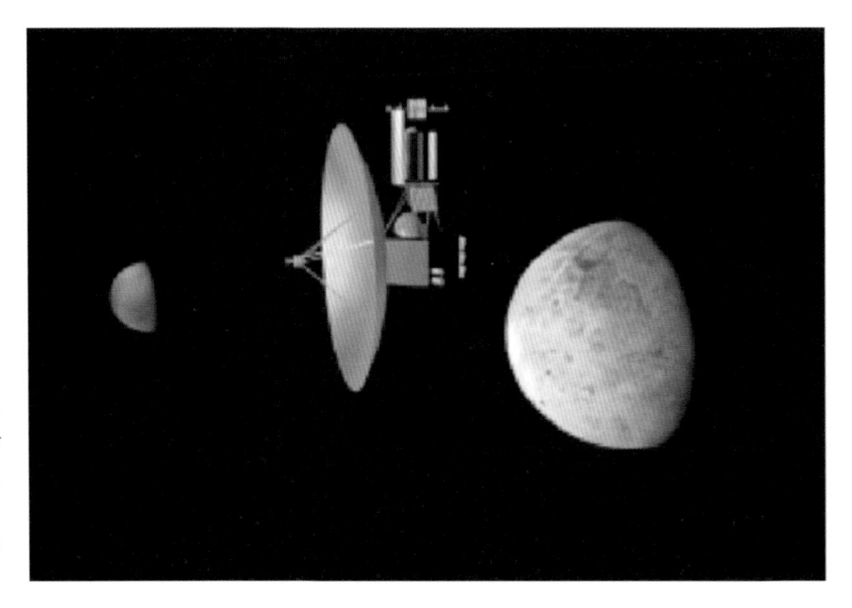

On the drawing boards is a daring mission called the Pluto-Kuiper Express. Its goal is to survey Pluto and Charon — and then sail on into the Kuiper Belt and visit one or more of the dis-tant cometary bodies that orbit beyond the realm of the planets. (NASA/Jet Propulsion Laboratory)

its surface area active. It is a Jupiter-family comet that orbits between 1.5 AU and 4.7 AU over a 5.5-year period. A major goal of the mission is to explore the structure of the upper layers of the comet's nucleus. By watching how the crater develops after the projectile hits, scientists will gain an insight into how many layers, if any, the comet has and their thicknesses. In addition, the impact will help determine whether its ices are tightly compressed or packed more loosely.

The explosion will also excavate material from deep within the comet that has not been exposed to space since the comet formed, more than 4 billion years ago. As these materials are heated by the impact and mixed together in the plume of vapor and ejecta, Deep Impact's instruments will investigate their chemical composition and relative abundances. Mission scientists chose to make the projectile of copper because it will vaporize in the impact and copper is easy to identify in the spectrum of the ejecta. Since copper is also a highly improbable ingredient for a comet, scientists are assuming that all the copper they detect comes from the projectile alone.

If the flight goes as planned, the launch will be in January 2004, an Earth flyby will boost the spacecraft's velocity a year later, and the two-part spacecraft will approach the comet on July 4, 2005, when it lies 83 million miles (134 million km) from Earth. The impacting projectile would separate from the instrument platform several hours before the hit and continue racing toward the comet, aiming for a point on its day side. A camera onboard the projectile would give scientists increasingly detailed looks at the target site as the projectile heads for impact. Meanwhile, the part of the spacecraft carrying the instruments would fire its thrusters to shift to a non-collision course and slow down a bit. At the time of impact, it would lie less than 300 miles (500 km) away from the comet.

The impact explosion is expected to be bright enough to be seen from Earth, even in amateur-size telescopes. Mission scientists plan to coordinate the spacecraft studies with simultaneous ones from observatories on Earth and in orbit to round out the picture. They plan to receive images of the impact from the spacecraft nearly in real-time, and these will be broadcast on television to the world.

A mission that almost borders on science fiction is the **Pluto-Kuiper Express**, a proposed JPL mission whose major focus lies on surveying Pluto and its moon Charon, the last planetary target that remains unvisited by spacecraft. Various scenarios for a mission have been developed in detail using different launch dates and flight times, although none has assured funding. The currently favored plan calls for launch in December 2004, a flyby of Jupiter in March 2006, and a flyby encounter with Pluto and Charon in December 2012. After surveying the two bodies, if all is still going well with the swiftly moving craft, it will be targeted to tour among the cometary objects in the Kuiper Belt (see Chapter 1), some 50 astronomical units from the Sun. Since communications with Earth from beyond Pluto would take many hours, the spacecraft will be equipped to make routine searches for Kuiper Belt objects on its own.

The mission would be a historic undertaking, similar in its way to exploring the last continents on Earth. But it poses real challenges in technology and timing, not least of which is a travel time lasting many years. A more critical timing problem concerns Pluto itself: its atmosphere is slowly freezing out in the form of frost and snow as Pluto's orbit carries it away from "perihelion summer." When Pluto's entire atmosphere collapses onto the surface sometime in the 2020s, it will take a significant part of the mission's scientific value along with it. To be worth the effort and expense, therefore, the Pluto-Kuiper Express must reach Pluto somewhat before then. Yet it remains to be seen if the budgetary and technological problems can be solved in time. And if no Pluto mission, then no Kuiper Belt tour either.

Finally, there is **Ulysses**, a mission in progress. Launched in October 1990, this joint project of the European Space Agency and NASA has been orbiting the Sun in a six-year-long trajectory that carries it over the solar poles. From this vantage point it studies the Sun's activity and its influence on the interplanetary environment.

Because comets, especially their ion tails, are sensitive tracers of the solar wind and its fluctuations, scientists with the Ulysses project established a Ulysses Comet Watch in 1992 and ran the program until 1998. The purpose of the program was to use amateur and professional observations of cometary gas tails — mostly photos showing the knots, kinks, and disconnection events — to indicate abrupt changes in the magnetic field embedded in the solar wind. Combined with spacecraft measurements, the observations let scientists reconstruct the electromagnetic activity of Sun and solar wind in much greater detail than any one spacecraft could show.

The Comet Watch revealed that the solar wind blows steadily and smoothly in the Sun's polar regions, its high-latitude velocity is around 460 miles per second (750 km/s) versus around 280 miles per second (450 km/s) at the equator, and disconnection events in a comet's gas tail occur only in the equatorial regions.

If comet missions got off to a slow start compared with other planetary probes, they have clearly made up for lost time. The next few years promise to bring a striking wealth of information for cometary science — both in regard to ordinary comets and to the Great Comets whose names echo through the history of astronomy and mankind.

6 Comets and Cultures

Comets and human beings have had a long and turbulent relationship, characterized mostly by bad feelings on our part. What prehistoric societies thought when they saw a Great Comet scribing a path in the sky will probably remain forever a mystery. But it's not hard to guess. In the eyes of everyone alive up until about 300 years ago, comets were bad news. The first references to comets appear in Chinese annals from the eleventh century BC, and right from the start they were regarded as heavenly portents, for good or ill, but overwhelmingly the latter.

That comets should carry a negative message is something that strikes most people in the developed world today as puzzling or downright bizarre. If we see a comet, we feel brushed by a sense of awe, we enjoy its beauty, and we may ponder the vast stretches of time, and the distances it has traveled, since its last visit. But fear? Dread? These don't really enter into it. When we stumble across a solemn pronouncement like Jacques Gaffarel's from 1650 — "If [comet tails] bear the figure of a sword, they presage desolations which shall be caused by the sword" — today's sensibility is more apt to see an overvivid imagination at work rather than a kind of National Weather Service advisory of impending severe political storms.

Surrounded by the lights of our cities and immersed in a self-preoccupied multimedia society, it's hard for us today to understand the effect of a bright comet on former cultures. We're too far removed in time, space, and mentality. To begin with, the appearance of a comet, even a Great Comet, may not seem all that stupendous to us. Few comets can compete on equal terms with the glare of billboards, streetlights, and store signs, let alone public spectacles such as fireworks or even movies with flashy special effects. Second, we have other pipelines to future events. Instead of worrying over the meaning of a comet and what it might portend, we leave prognostication to pundits and stock brokers. If there is any echo now of yesteryear's view, we find it only in the super-

The Aztecs of Mexico noted the long filmy tails of comets and called them "smoking stars." Believing the arrival of the Spaniards was a fulfillment of prophecies brought by recent comets, the ruler Moctezuma felt that the heavens had turned against him. This belief helped undermine his resistance to the Spanish invasion. (Ruth S. Freitag, Library of Congress)

market-tabloid headlines, and who takes those as reliable and serious statements of fact?

Should we wish to recapture the ancient sensibility, we might begin by imagining how the night sky appeared to those who lived thousands of years ago. It was a time when the night itself was profoundly dark, unless the Moon were visible. Nights were filled with justifiable fears of predatory animals and of humans no less dangerous. The only artificial light came from campfires, torches, and oil lamps. Seen from some open hilltop or megalithic monument, the stars paraded overhead every clear night, slowly shifting with the seasons. Nobody now catches a glimpse of this majestic sight except perhaps during vacation in the

Good news, bad news. In 1066, a year with a grand visit by Halley's Comet, Harold, the ruler of Saxon England, learns that his kingdom has been invaded by William of Normandy. Many in England took the comet as a bad omen — but only after Harold was killed that year in the Battle of Hastings. William naturally claimed the comet had forecast his successful conquest. (From the Bayeux tapestry; by special permission of the city of Bayeux)

North Woods or if they live far away from a city. Yet the sky's ancient power hasn't vanished completely. When you talk to someone who has just returned from such a trip, you can hear in their voices the wonder of what they've seen.

The ancients were a lot more familiar with the night sky than we are. They knew the stars and their patterns intimately and told elaborate stories about them to each other and to their children. For cultures without written records, all of a people's learning and experiences must be carried in their memories and handed on through tales and legends. The slowly shifting constellations made a kind of storybook in the sky.

The ancients saw that most of the stars stayed in place relative to each other. But among these "fixed stars" drifted a handful of wandering ones — the planets — that moved from week to week relative to the starry backdrop. Planets also brightened and dimmed in ways no fixed

By the nineteenth century, comets had become spectacles to be enjoyed by a literate and knowledgeable public rather than terror-laden portents. The 1835 apparition of Comet Halley — a fine one — was the second predicted return for the comet, and it attracted much interest from scientists and the general public. (From a French broadside pamphlet; International Halley Watch, NASA/JPL)

star did. Since planets moved and changed much as living things do, it was natural to suppose that they carried a link to this world and perhaps could influence its inhabitants. Out of this arose astrology. Given the eternal human hunger to find order in a risky universe, it helped make the world a bit less frightening.

Now, if one believes that slowly moving planets hold power, consider what bright comets embody. Their appearances are rare, unpredictable, and often dramatic. Comets may bear an uncomfortable resemblance to blazing fires or daggers or sword blades. They move among the fixed stars more quickly than any planet, they can shine much brighter and, most important, they appear to flout the rules controlling the movement of celestial bodies. Even prehistory's skywatchers must have seen that planets were always found in a band of constellations through which the Sun traveled during the year. They would take this to mean that somehow the Sun held sway over the wandering stars and kept them in order. But comets followed to no such circumspect course. They wandered the

The Great March Comet of 1843 became bright enough at one point to be seen during daylight close to the Sun. Astronomers now know it to be one of the famed Kreutz family of sungrazers. Here its tail stretches across the evening sky of Paris below the feet of Orion. (Ruth S. Freitag, Library of Congress)

sky at will, coming and going with no regard for the predictabilities of planetary motion. Such arrant waywardness proclaimed that the arrival of a comet was a special event. It was literally a "heads up!" shout that something significant was afoot in the heavens.

The earliest civilizations of the Middle East bequeathed few astronomical writings on comets. Yet, as soon as comet references do appear, the impression they make is almost universally as harbingers of doom and misfortune. Now and then a comet was thought to presage the birth of a good ruler, but most comets simply heralded bad times for rulers and ruled alike. This theme continued in the speculations about the natural world by ancient Greek philosophers. Their earliest writings on comets survive mostly in fragmentary quotations assembled by later writers. Yet the overall gist is clear — comets foretoken disasters.

The only ancient Greek writer whose cometary ideas we can study in detail is Aristotle (384–322 BC). He believed that comets did not belong to the heavens proper. They were "meteors" — his term for phenomena that occur within the atmosphere (hence our word meteorology). Rain, clouds, auroras, high winds, and shooting stars are also meteors in Aristotle's sense. He thought that comets occurred when dry explosive vapors rose from the Earth to the highest part of the atmosphere. There they would be ignited by friction from the movement of the first of the heavenly spheres. These surrounded Earth like the layers of an onion and carried the Sun, Moon, and planets. When the dry vapor caught fire slowly, the result was a comet. (When the vapor caught fire quickly, it made a shooting star.) For Aristotle, cometary "exhalations" were signs that the world was in disorder. Terrifying disasters were much more likely to follow whenever strong winds and earthquakes occur, the latter being the source for the vapors that become comets.

The Roman astrological poet Marcus Manilius, writing between AD 9 and 15, echoed this view straightforwardly. "In times of great upheaval... comets blaze forth into life and perish." But his tone is far more lurid than Aristotle's. Throughout Manilius' cometary writings, images of hair (considered unlucky) and fire and flame mingle with those of destruction and corruption. "The fire of comets may match with swollen flames casks with greatly distended paunches... Death comes with those celestial torches, which threaten earth with the blaze of pyres unceasing."

It's not hard to hear in those extravagant outpourings echoes of the biblical book of Revelation. But it's pointless to ask of Manilius what comets *are*. He didn't care about their physical nature — it was their significance for prognostication that energized him: "It may well be that by means of these moods and conflagrations of the sky, heaven in pity is sending upon Earth tokens of impending doom..."

The Roman writer and philosopher Lucius Annaeus Seneca (5 BC?–AD 65) was a contemporary of Manilius and acknowledged this craving for meaning in a comet's appearance. He wrote,"If a rare fire, and one of unusual shape, appears, everyone wants to know what it is... and whether he ought to admire or fear it." Yet if Seneca saw a role for comets as omens, he was unimpressed by what earlier writers had said about them. Despite his claim that debunking other philosophers is like

In 1910, the return of Halley's Comet became the world's first "media event," generating a global interest that spread into many aspects of daily life. (Observatories of the Carnegie Institution of Washington; May 8, 1910)

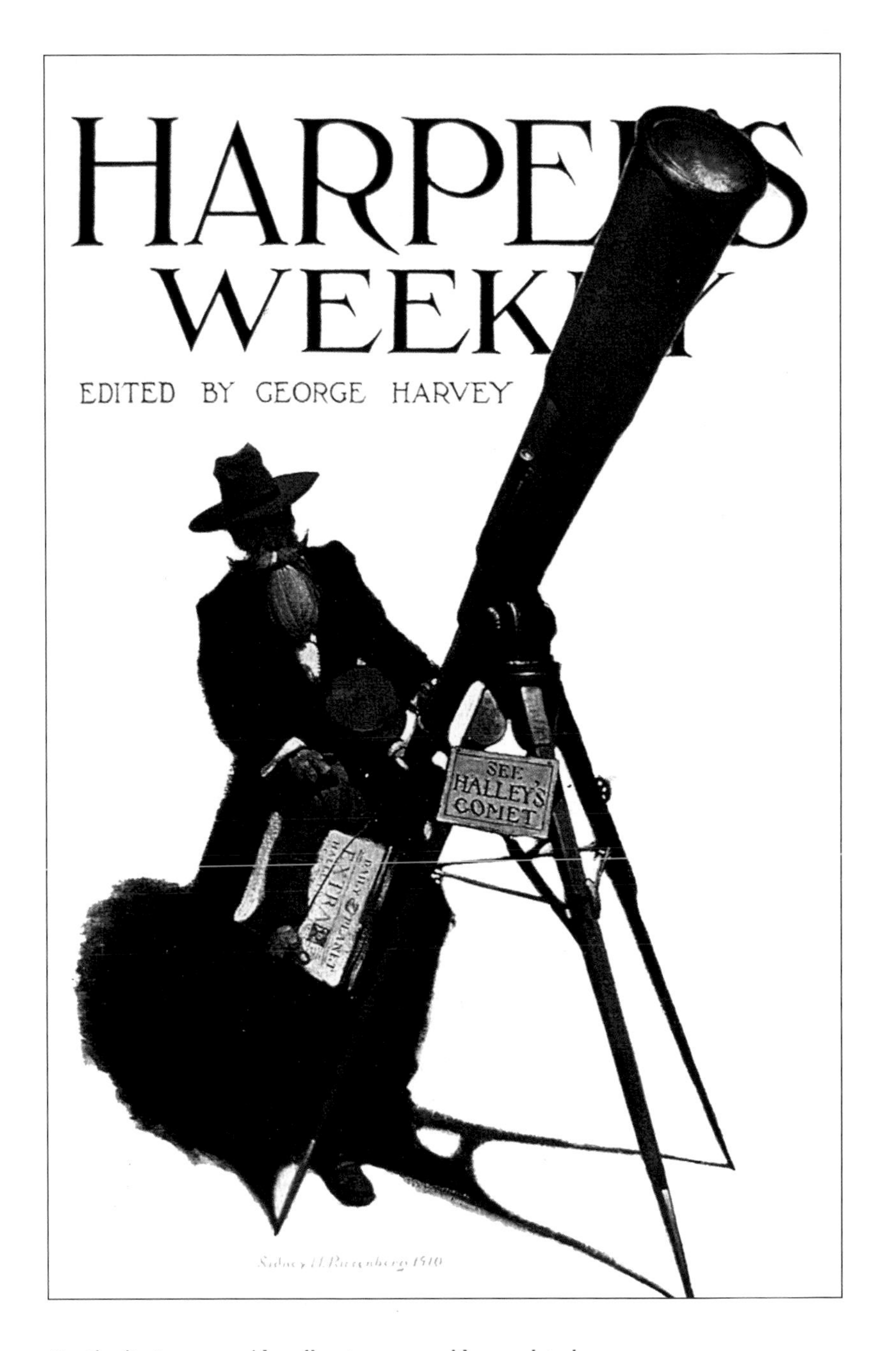

Continuity. In 1910 a sidewalk astronomer old enough to have caught Halley's Comet on its 1835 visit shows it off to a newsboy young enough to see it return in 1986. (Painting by Sidney Riesenberg; Ruth S. Freitag, Library of Congress)

Halleymania. Halley's 1910 visit brought out cartoonists who had loads of fun with the mass excitement gripping society. (Illustration by W. Heath Robinson; Ruth S. Freitag, Library of Congress)

Ballyhoo and Halley, too. Every product that could reasonably be linked to the comet, was — along with some that took real effort on the part of the ad copy writer. (Ruth S. Freitag, Library of Congress)

punching the wind, Seneca went to considerable length — too long to be succinctly quoted here, unfortunately — to restate what they said about comets and to dissect their arguments in detail before presenting his own views.

His views followed a pretty skeptical path since Seneca was dedicated to finding natural explanations for comets. Basically, Seneca thought comets were more show than significance. They come and they go. What makes them so eye-catching, he said, is merely their unusual appearance, which sets them apart from other celestial phenomena. And this is just what you'd expect, he argued, since ordinary sights draw little notice — "no one watches the moon if it is not in eclipse." For his part, he thought that focusing on comets' role as portents overlooks what is truly wonderful about the heavens: the everyday events that occur. "People ask again and again and want to know whether [a comet] is an omen or a star." Such people, he said, miss the point. "By Hercules, no one could study anything more magnificent or learn anything more useful than the nature of the stars and planets..."

For Seneca, comets just weren't as important as public hysteria made them out to be. They were simply natural objects, and part of the beautiful diversity of phenomena to be seen and, perhaps, understood. "Some day there will be a man who will show in what regions comets have their orbit, why they travel so remote from other celestial bodies, how large they are and what sort they are."

What seems most remarkable to us about Seneca is that his ideas are so ordinary and grounded in everyday experiences. There's none of the overheated mumbo-jumbo that make Manilius and many others read like an ancient *Tales from the Crypt*. Seneca's thinking sticks closely to visible phenomena and ordinary analogies, such as terrestrial winds and hearth-fires and other commonplaces. After reading him, you come away impressed. His understanding of comets naturally misses the mark in light of our better knowledge 2,000 years on, yet his attempts to reason them through strike a surprisingly modern note.

Slightly later than Seneca, however, another Roman natural philosopher, Pliny the Elder (AD 23–79), fell back on the traditional Aritotlelian view. For Pliny there was a direct causal link between disasters and comets: "I hold that... [the comets] took place because the misfortunes were going to occur." His thinking was that disorders on Earth and signs of disorder in the sky — that is, comets and other phenomena such as auroras and shooting stars — come about because conditions in the universe are out of joint. When the universe is disordered, the heavens break out in comets just as someone sick with the measles develops red spots.

Around AD 150, an astronomer in Alexandria named Claudius Ptolemaeus (known to us as Ptolemy) wrote a magnificent book that summarized the astronomical knowledge of the ancient world. Ptolemy (AD 100? – 175?) provided an exhaustive mathematical description of the movements of the Sun, Moon, stars, and planets, all based on the Earth-centered model of the universe that had been current for centuries. The work is best known by the title — the *Almagest* — that it received in a later Arabic translation, and it virtually codified all of astronomy for nearly 1,500 years.

For Ptolemy, however, comets were *not* part of the heavens so he omitted them entirely from the *Almagest* and dealt with them only in his astrological writings. In Ptolemy's eyes, comets had value only as portents. The region of sky where they appeared (and the direction where the tail pointed) indicated what part of Earth would be affected. Comets exerted their ill effects for as long as they were visible. Also, Ptolemy apparently believed that comets behaved rather like the old saying "Red Sun in the morning / sailor take warning." He wrote that comets seen in the east near sunrise gave a strong indication of imminent danger, while the threat from one seen in the west was less immediate.

As classical antiquity ended, Christianity began to dominate all learning in Europe and in the lands around the Mediterranean. The gloomy view of comets continued unabated. For the Church, Aristotle's physics (and Ptolemy's astronomy) provided the framework for all studies regarding nature. At the same time, the political and, later, the religious fragmentation of Europe and the resulting conflicts made life precarious for ordinary people. Thus it was only natural that the portent-laden image of comets should dominate Western astronomy for more than 14 centuries after Ptolemy — until well into the 1600s. During those ages of fervent Christianity, thinkers having a mystical bent often linked comets to other signs of a disordered universe: strange happenings and

1957 saw two bright comets in succession, Arend-Roland (seen here) in the spring and Mrkos in the late summer (see page 55). Arend-Roland was the first brilliant comet to be visible from the northern hemisphere since Halley's 1910 visit. (Elizabeth Roemer; US Naval Observatory, Flagstaff Station; May 3, 1957)

freaks of nature such as births of multiple-headed calves, showers of bloody or milky rain, and thunder booming out of clear skies. In their apocalyptic visions these signaled that bad times were on the way, and comets and the other prodigies were just God's way of letting us know.

Typical of such views are the sentiments recorded in a Venetian chronicle for AD 1006: "And so at the same time, a comet, the sign of which always announces human shame, appeared in the southern regions, which was followed by a great pestilence..." The chronicler in question is actually describing the appearance of a supernova, an

In the winter and spring of 1962, Comet Seki-Lines traced a bright course moving northwest out of Puppis. But it was a star performer mainly for the Southern Hemisphere. (Elizabeth Roemer; US Naval Observatory, Flagstaff Station; April 11, 1962)

exploding star with no physical relation to comets. But, while of course his astronomy knew nothing of exploding stars, his feelings about what comets mean are quite clear. Attitudes like this continued right into the Renaissance, and were found at all levels of society.

In eastern Asia, a similar pattern held. Chinese astronomers closely observed comets — which they called "bushy" stars and "broom" stars and other descriptive terms — for over 3,000 years. They did this mainly with an eye for determining the fates of rulers and dynasties. In

Chinese political philosophy, the emperor was both the Son of Heaven and its direct representative on Earth. (China wasn't known as the Celestial Empire for nothing.) Since the cosmos was tightly linked to terrestrial power, Chinese astronomy bore an official status in ways that held few parallels in the West. Astronomers were called upon to interpret the meaning of celestial events and put the right spin (as we might say today) on the emperor's doings. But this was often a precarious business given the intrigues of court factions, and the life of an astronomer could not have been a secure one, especially in times of troubles.

A silk-paged book on sky phenomena was found in a Chinese tomb in 1973. The book dates from 168 BC, although its text was probably composed several centuries earlier and codifies long-standing notions of the meaning of sky phenomena. Among its illustrations of rainbows, clouds, and suchlike are about two dozen drawings of comets, each showing a different shape of tail. The drawings are annotated with short descriptions that tell the kind of omen each comet portends. For instance, a comet with one type of tail indicates that a big harvest will follow, but also a civil war. Another foretells of a revolt in the army, while a third says that kings will die. On and on it goes. What is most striking about the book is that essentially none of the portents are good. If books of omens tend to mirror in their attitudes the prevailing beliefs of the society they come from, then the state of affairs in early China was no more settled or stable than the West, for all its cultural and technological advances, which at that point far outstripped Europe's.

While China's astronomical annals universally revolve around divination, the records have nonetheless proven highly useful in a scientific sense. The emperor's astronomers were charged with noting when a comet appeared, how long it remained visible, how lengthy its tail was and what it looked like, and where on the sky the comet traveled. Out of such meticulous records came the famous account of the Crab Nebula supernova that exploded in July 1054. Observations of the decline of this "guest star" helped astrophysicists draw a detailed picture of the exploding star whose violent death throes are still expanding into space. Likewise, Chinese comet observations let scientists trace returns of Halley's Comet back through 30 consecutive appearances to 240 BC.

The Great Comet of 1965. Comet Ikeya-Seki arrived in an era when many people were caught up in the burgeoning Space Race. Unfortunately, it reached greatest brightness in the dawn sky, so fewer people saw it than might have otherwise. (Richard Berry; October 1965)

If Kohoutek failed to become a Great Comet for the public, no planetary scientist was disappointed by it. The comet provided the first-ever detections of methyl cyanide and hydrogen cyanide in comets as well as the identification of silicon grains in the comet's dust tail. (Robert G. Roosen, John C. Brandt, Joint Observatory for Cometary Research — NASA Goddard Space Flight Center and New Mexico Insititute of Mining and Technology; January 13, 1974)

Similarly, an observation from AD 1106 may denote the comet whose breakup created the Kreutz family of sungrazing comets (see Chapter 2). The fragments of this shattered and scattered group include Great Comet Ikeya-Seki seen in October 1965.

Western astronomy once ignored Chinese records but the tide has long since turned. It's true that Chinese theoretical and mathematical astronomy never broke free of its astrological connections (unlike that in the West); thus it remains mostly a curiosity for specialists. But taken simply as observations, Chinese records of comets surpass or match any in the West until well into the European Renaissance. It was only afterwards that Western astronomy began to stride ahead in instrumental techniques, observational precision, and physical understanding.

The interaction between comets and Western culture remained thoroughly dismal and gloomy into the seventeenth century, the time of Isaac Newton and Edmond Halley. The change began first in England and soon spread to France and other countries where the Scientific Revolution was drawing allegiance among the learned. The discovery by Newton, Halley, and others that comets move according to mathematical laws presented a picture of an orderly clockwork universe that stood in marked contrast to the apocalyptic visions clung to by the uneducated. And even if (as some of the learned still believed) a comet was a warning from an angry God, once a comet appeared, astronomers were able to show that it behaved in ways which mathematics could fathom. This had the effect of demoting comets from capricious heavenly omens to a kind of planet that might arrive unheralded but which moved in predictable ways.

At the same time, however, Newton and his contemporaries knew perfectly well that if comets move among the planets, on occasion a comet will hit a planet and the result must surely be a catastrophe. (In fact, Newton believed that a constant infall of comets kept the Sun's fires burning.) However, this scientific catastrophism met a curious fate that effectively banished it from serious discussion for several hundred years.

As astronomy historian Sara Schechner Genuth writes, in the time of Newton and Halley, England was undergoing political upheaval and had just emerged from a bloody civil war. Many would-be revolutionaries either believed in cometary and astrological prophesy or used it for ulterior political ends that involved overturning the established order of society. As a result, the learned and upper classes (virtually one and the same) gave acceptance to the mathematical side of Newton's discoveries — calculating orbits and the like — but did not acknowledge catastrophic events, as these seemed to imply a support for radical political change. Ignored, dismissed, and suppressed, catastrophes virtually disappeared from scientific debate. A like reaction ensued across the Channel in France and other countries where scientists dwelt within similar social and political structures. (In the next chapter, we'll pick up the story of cometary catastrophism and see how it reemerged in the last few decades.)

During the eighteenth century the clockwork universe gradually seeped into the popular culture of Europe and the Americas. As it

Comet West made a splendid sight in the predawn sky in the spring of 1976, yet attracted fairly little public attention. This was probably due to the time of night when it was visible and astronomers being publicity-shy after the Kohoutek affair two years earlier. (Steve Padilla and Ron Royer; March 13, 1976)

spread, it supplanted the fearful and doom-ridden views held by many, especially among the lower classes. By the nineteenth century, a bright comet had long since ceased being something whose meaning had to be divined for the good of humanity. Instead, it was seen by astronomers as an opportunity for study and by the public as a spectacle to be enjoyed. The many Great Comets of the 1800s found eager audiences in the growing middle classes, who were more literate, better educated, and had a taste for scientific amusements. In this spirit, people marveled over nineteenth-century Great Comets such as those of 1811,

1835 (an apparition of Comet Halley), 1843, 1858 (Donati's Comet), 1861, and 1882.

And then came Comet Halley in 1910. The reappearance of this famous comet according to astronomers' predictions coincided with the growth of large-circulation newspapers and periodicals and the development of mass advertising. In the twilight of the Edwardian era — in fact, King Edward VII of England died two weeks before Halley made its closest approach to Earth, and his funeral was held on the very day of it — the vast publicity that surrounded the comet's arrival far exceeded anything it had received before. Only the recent visit of Comet Hale-Bopp could be said to rival it. The appearance of Halley's Comet in 1910 was the world's first international media event, and it was outstanding.

It started normally enough. Astronomers knew where to look for the incoming comet, and Max Wolf of the Heidelberg Observatory in Germany was the first to photograph it on September 11, 1909, approaching just about on schedule. It was first spotted by eye a few days later by Sherburne W. Burnham (no relation of mine, incidentally) using the great 40-inch refractor at Yerkes Observatory in southern Wisconsin. Soon every observatory on Earth had the comet in its sights and was studying the growing activity as the comet approached. Articles in newspapers and magazines began to pepper readers with information about comets, as well as directions on how best to see it. Regular doses of Halleyana became a feature of public discussion.

Observatories weren't the only places staying up late, fixated on the comet. Editorial and satirical cartoonists used the comet's arrival as a metaphor (or running joke) for whatever their particular issue was. Music publishers raced to get out sheet music for the thousands of pianos in parlors across the land, all with tunes carrying titles like the "Halley's Comet Rag." Likewise, mass advertisers attached comet images to every product in the world. Soap, coffee, custard, corsets, perfume, beer, schnapps, umbrellas, and furniture were only a few of the ballyhooed items. Sales of telescopes and binoculars shot through the roof. Astute purchasers moved quickly, so well-made pieces of equipment were bought up early. In the end any old piece of junk with a tube and a couple of lenses could be sold for an absurd price to a desperate and uncritical public. (Don't smile; exactly the same thing reoccurred at Halley's most recent return in 1985–86.)

Then a surprise happened. In early 1910 a bright new comet burst into the growing apparition of Comet Halley. The new comet was discovered by gold miners in South Africa on January 13, and because the news took several days to be confirmed it never received the name of any particular discoverer. History simply records it as the Great Daylight Comet or the Great January Comet. It reached greatest brightness (about equal to Venus) in the evening sky after January 18, the date of perihelion. But it disappeared quickly from naked-eye view, being gone by the end of the month. The Great January Comet was notable for its long tail, reportedly reaching 50° at maximum. Vivid and totally unexpected, the Great January Comet made a strong impression on a public already well primed for comets. (In truth, going by details in their accounts, it is this comet that some surviving witnesses of Halley in 1910 are actually remembering.)

It's not clear just when and where people started to panic. By 1910 astronomers had known for several years that comet tails contain rarefied gases, including molecules such as CN, better known in its stable laboratory form (C_2N_2) as cyanogen, a deadly poison. On February 6, 1910, observations confirmed that Halley's gas tail (like that of many other comets) contained cyanogen. A few days later, noted American comet-hunter and astronomer E. E. Barnard comfortably announced that "there was great reason to hope" that the comet would have "plenty of tail to reach to Earth" on May 18, the date when calculations showed that Earth might pass through Halley's vaporous tail. Barnard may have viewed the prospect with equanimity, but newspapers all over the world instantly linked the facts together and pressed leading astronomers to comment on the possible danger. Most of them sensibly replied that the amount of cyanogen in Halley's tail was much too little to have any detectable ill effects. Several astronomers even mentioned that Earth had already passed through comet tails before, harmlessly. But not all experts were so reassuring.

Henri Deslandres, director of the Meudon Observatory near Paris, agreed that the tail's contact would probably do no harm — but then he emphasized that the tail contained enormous quanitities of cyanogen.

If Halley wasn't very great from the Northern Hemisphere in 1986, the Southern Hemisphere had a much better view. This image, taken from Easter Island, shows ionized gas streamers rippling in the solar wind. (William Liller; Celestron Schmidt camera, 4 minutes on Fuji 400 print film; International Halley Watch; March 8, 1986)

The springtime wonder — Comet Hyakutake — sped across the northern sky in March 1996, scribing a track in public awareness that Comet Hale-Bopp was to follow a year later. (David Churchill; 200 mm f/2.8 lens, 30 minutes on hypered Fuji 800 print film; March 23, 1996)

Much worse, the famed astronomer and popular science writer Camille flammarion said that the cyanogen could infiltrate the atmosphere and snuff out all life! Knowingly or not, Flammarion was presenting as fact what amounts to science fiction. In the charged media environment, even the sober periodical *Scientific American* felt compelled to address the fears. Acknowledging that a comet might cause calamities if it struck Earth, the magazine carefully explained that Halley's nucleus would not come close and its tail would not hurt anyone. But it's unclear if this reassured the general public.

The problem was that the public was receiving mixed messages: one professor would be quoted as saying the risk was real, while another would scoff. Faced with experts who couldn't agree, many members of the public figured it was better to be safe (and scared) than sorry.

So as the predicted time approached — the night of May 18 to 19, 1910 — those who were prone to panic sealed house chimneys and windows and stuffed rags under the doors of bedrooms. Murderers, figuring it was all over anyway, confessed their misdeeds. It appears that a number of people actually committed suicide in anticipation of the expected general doom — although the famed story of the virgin being rescued when she was on the point of becoming a human sacrifice was a hoax. Citizens in many cities hoarded bottles of oxygen, an act Pope Pius X condemned. Hucksters sold phony "comet pills" to ward off the ill effects. Churches saw record attendances, and rural people restocked and prepared their storm cellars. In every large city on the night itself, people who weren't barricaded in air-tight rooms flocked outdoors to rooftops, to streets, restaurants, cafés, and parks in spontaneous mass gatherings. Many were nervously jovial as they waited up through the night for — well, who could say just *what* would happen?

What finally happened, of course, was that the night, like all others, ended as the Sun rose. The comet's tail appeared to have missed Earth by an undetermined distance. Everyone drifted home to have a good "comet breakfast" and a chuckle over the whole thing, spiced with a twinge of relief. Even if Earth had passed through the tail, the amount of cyanogen falling into the atmosphere would have been negligible, just as most astronomers had said. And yet their fears of asphyxiation were perfectly justified, in fact. People actually were being poisoned — by the air pollutants (coal smoke mostly) that lay in a thick pall over every industrialized city, choking its inhabitants. But comets had nothing to do with that.

After Halley's 1910 visit, Great Comets took a vacation for a number of decades. Several fairly bright comets came and went including two fine ones in 1957, Arend-Roland and Mrkos. But none was truly spectacular and certainly none gripped the general public's imagination as Halley had.

Moreover, the passing years saw changes that have affected how all comets, great and otherwise, are perceived by the public. Since the early years of the century, there has been an important shift in the way the media convey information about events, including celestial ones such as comets. For example, at Halley's return in 1910, everyone looked to the newspapers for the latest information, but it was delivered mostly in words. Images showing the comet were usually diagrams — pieces of artwork, drawn or painted. Astronomers were photographing the comet, but for technical reasons these were hard to reproduce and relatively few saw print. This meant that for people in 1910 virtually the only way to experience the famous comet was to go and look at it first-hand. You had to step outdoors and see it personally. And the image of the comet that you carried in your mind derived mainly from what your own two eyes had shown you.

Things have changed a lot since then. The growth of mass media and especially their ability to deliver pictures — first in print, then via television, and now with the Internet — means that today we are inundated with a torrent of images. This in turn means that any first-hand eyewitness impressions we form of a comet are inescapably overlain with photos and images of it drawn from observatories, spacecraft, and backyard astrophotographers all over the world. And many of these are far more visually striking than what we see on our own. The result is that our personal views no longer have the unchallenged primacy they once did.

There's no point now in asking if the older way was better. It was simply different. (After all, no one is going to uninvent television and computers.) I think there's no question that we've gained by the changes. While individual impressions of a comet may now be somewhat diluted by an (over)exposure to hordes of images from other people, the tradeoff seems acceptable. If our memories of a Great Comet are now less wholly "real" and personal, they are undeniably richer, more varied, and more detailed.

In the lifetimes of most people today, the first Great Comets they remember hearing about are usually Comet Ikeya-Seki of 1965 or Comet West of 1976. (The much-heralded Comet Kohoutek of 1973–74, of course, failed to become a Great Comet.) Both Ikeya-Seki and West, however, made comparatively little impact on the public. They appeared when skies had become bright with city lights (as they still are) and, what's probably more relevant, the comets shone their best in the wee hours before dawn when relatively few people are stirring.

On the other hand, both Ikeya-Seki and West arrived in an era when the state of comet science was far in advance of that which greeted Halley in 1910. Intellectually, at least, mankind was living in a new solar system, thanks to spacecraft sent out from Earth. During the Space Age, planets changed from being remote dots of light swimming in a telescope's eyepiece to worlds with geographies and geological histories. One of them — our own Moon — has now been walked upon, and several hundred pounds of its rocks sit in laboratories where planetary geologists use them to piece together the skein of lunar history.

As explained in Chapter 5, the practicalities of spaceflight usually prevent scientists from sending off missions to greet bright, long-period comets on their home turf. Flights out to the Kuiper Belt, let alone the Oort Cloud, lie a decade or more in the future. But all comets with orbits passing through the inner solar system are fair game, and the return of Comet Halley in 1985–86 saw the first concentrated efforts to aim space probes at a comet. The results were highly successful — and a good thing too because, for most viewers on Earth, this visit from Halley wasn't very thrilling. The comet remained relatively far from Earth (and near the Sun in the sky) when it was brightest, and moonlight also interfered even when Halley could be seen clearly. Moreover, best views came from the Southern Hemisphere, half a globe away from the world's main centers of population. The result wasn't exactly a fiasco — Halley is too famous for that — but the apparition was definitely underwhelming for most people, who naturally won't live to see it again.

After Halley's 1986 visit slipped into the history books, it took the discovery of Hale-Bopp in July 1995 to give everyone new hope of seeing another Great Comet in their lifetimes. Then lightning struck a second time.

Just as the Great January Comet of 1910 interrupted the smooth build-up of interest in Comet Halley that year, so Comet Hyakutake broke into the long arrival of Hale-Bopp. Almost exactly one year before Hale-Bopp would reach its brightest, Comet Hyakutake appeared, rapidly brightened to become an impressive sight even for ordinary, city-bound people, and then disappeared quickly. The whole apparition, however, ran its course too swiftly for the comet to make a big impact on the general public.

While many people have their own Hale-Bopp experiences to cherish in memory, thanks to the media, millions of people around the world became caught up in one particularly bizarre story. It began with an amateur astronomer's hasty mistake and ended in a collective suicide by members of a religious group in California.

On November 14, 1996, Chuck Shramek, an amateur astronomer from Houston took a photo through his telescope of Comet Hale-Bopp using a CCD (electronic) camera. The photo showed the comet and a fuzzy object nearby. Because the fuzzy object appeared elongated somewhat like the planet Saturn when seen poorly focused, Shramek called it a "Saturn-like object," or SLO for short. He observed the SLO for more than an hour, believing that it was following the comet. Shramek wanted very much to know what the SLO was, but it was not identified in the computer sky-viewing program (MegaStar) that he was using as an observing guide. When MegaStar came up with nothing, Shramek leapt to the conclusion that the SLO was something entirely new and unknown to science. Thinking that he had a major astronomical discovery on his hands — but without asking any other observers to verify the object — that same evening Shramek phoned a national radio talk-show run by Art Bell that specializes in sensational subjects. Almost immediately Shramek was on the air, informing all who were listening about a mysterious object out in space following Comet Hale-Bopp.

Equally quickly, it seems, people with a strong belief in the paranormal were saying that Shramek's SLO was a huge spaceship, supposedly three times bigger than Earth. A organization promoting extrasensory powers (the Farsight Institute) quickly got on the bandwagon, claiming to be able to remotely view inside the SLO, apparently just by thinking really hard about it. The Farsight Institute then breathlessly announced, to no one's surprise, that it was indeed a spaceship and it was inhabited. Meanwhile, the publication of Shramek's original photo on the Internet let professional and amateur astronomers take a good look at it. They soon identified the "SLO" as an ordinary 9th-magnitude star designated SAO 141894. The star had not been identified by Shramek's software because of a bug in the design of the program's display. The star's abnormal shape in the image was caused by the telescope's secondary mirror and the nature of the CCD chip used in place of film by Shramek's electronic camera. And its "motion" was due to an eager imagination on Shramek's part.

When the correct explanation came out — and the error had been reproduced by others using the same software — the issue should have quietly disappeared. What actually happened, of course, is that people dedicated to seeing conspiracies everywhere continued to believe in the SLO-spaceship and dismissed the explanation as just a smokescreen for some kind of government coverup. As proof, they cited other photos taken later, which they claimed also showed the ship following the comet. Unfortunately for conspiracy-mongers, however, all of these subsequent photos turned out to be crude fakes, cooked up by hoaxers using the original starfield photo and image-processing software. Astronomers exposed the fakes as fast as they surfaced on the Internet, but the SLO true believers went on believing.

The farce turned serious in March 1997 as Comet Hale-Bopp was nearing its closest approach to Earth. In a suburb of San Diego, California, 39 members of a religious group called Heaven's Gate were found dead in the mansion that served as their home and office. It was a carefully organized mass suicide, complete with a video-taped message from the group's members and leader. The group described its beliefs in phrases that blended New Age philosophy with aspects of Christianity. The Heaven's Gate members said that Hale-Bopp's arrival was a sign that a spaceship had come to take them to a higher level of mental development beyond humankind. It was time, they said, to leave their

Heaven's Gate. As Comet Hale-Bopp approached Earth in March 1997, 39 members of this California religious cult committed mass suicide. Considering that the group earned a living by designing Internet home pages, it was ironic that they took their lives just as the Internet was creating a new mass medium to bring Great Comets to the public.

earthly shells behind and become creatures of pure spirit. "Hale-Bopp's approach," they wrote, "is the 'marker' we've been waiting for — the time for the arrival of the spacecraft from the Level Above Human to take us home to 'Their World' — in the literal Heavens. Our 22 years of classroom here on planet Earth is finally coming to conclusion — 'graduation' from the Human Evolutionary Level. We are happily prepared to leave 'this world'. . . ."

The circumstances of the mass suicide and its link to the comet produced a palpable shock in the public mind. It was reinforced a month later when two remaining cult members also attempted suicide, one successfully. Whatever people thought of the group's action, it made a statement that came straight at the world out of an earlier age. For the cult members at least, the arrival of Comet Hale-Bopp was no ordinary natural event. Instead, it was a signal from heaven, deeply fraught with theological meaning. This notion would have resonated strongly with

Almost no one in the general public realized it, but the tracks of Comets Hyakutake and Hale-Bopp passed near the same stars exactly one year apart. On the left is Hyakutake on April 9, 1996, while on the right is Hale-Bopp on April 9, 1997.

anyone living before the Renaissance, yet it has all but vanished from Western society today.

The arrival of Hale-Bopp also heralded another change, one as modern as could be. For the public in the developed world, Hale-Bopp and Hyakutake were the first Great Comets of the Internet. This worldwide linkage of computers was a research tool that had been used by scientists for a generation. For the average home computer owner, however, the Internet really started to blossom during 1995 and 1996, around the time Hale-Bopp and Hyakutake were discovered. Previous comet apparitions had been events that people followed by reading magazines or newspapers or (in recent years) on radio and TV. Almost overnight, it seemed, the new development allowed anyone in the world whose computer was linked to the World Wide Web — at home, work, or school — to follow comet discoveries, swap gossip and observations by e-mail, be alerted to new phenomena, view thousands of photos electronically, and so on. As it became a new participatory public medium, the Internet made the whole experience of these two comets vastly richer for many people, including some who might otherwise have paid a comet little attention.

About three centuries ago, just yesterday in the history of mankind, the Scientific Revolution began to alter the way our culture looks at comets. From being dreaded signs of an angry heaven, comets gradually changed to become public spectacles — celestial light-shows put on by nature for the pleasure of us all.

Yet this shift is still relatively new and on occasion darker thoughts intrude since they seldom lie far below the surface. It's unlikely that the prickle of ancient awe that a Great Comet can still raise in us will disappear any time soon, even as civilization brightens the nights and our greater knowledge brings more understanding. Such ideas are very deeply ingrained in human beings and link us with the earliest sky-watchers of our kind.

Nor is the disquiet totally unfounded. When astronomers of the seventeenth century realized that comets were ordinary members of the solar system, the discovery robbed comets of their supernatural threat while it gave them a physical one. After all, if comets wend their way among the planets, it takes no imagination whatever to see that now and then they must surely hit Earth and cause catastrophic destruction. We turn to this changed sense of cometary danger — grounded in a physical standpoint rather than a theological one — in the next chapter.

7 Danger From the Sky

In the previous chapter we saw that one highly disturbing implication of the work of Newton and Halley — that comets may strike the Earth from time to time — was largely ignored by scientists for the past 300 years. During the seventeenth century, scientists had specific reasons (rooted in contemporary politics) for turning their backs on the idea of cometary catastrophes. But exploring in detail why this idea continued to face opposition from later scientists would require a much larger book than this. The important point is that, by the early nineteenth century, astronomy (and the fields of geology and biology) had each adopted a view that change in nature occurs very slowly and involves only the processes visibly at work in the world around us. In such a climate of scientific opinion, ideas hinting that catastrophes might also play a role in change were distinctly unwelcome.

To oversimplify, but not by much, it took the Space Age to force a change in perspective. In the 1960s, planetary scientists started to survey the worlds around us using spacecraft. They discovered that by an overwhelming margin the most common feature in the solar system is the impact crater. Entire planets, moons, and asteroids turned out to be covered with them, at all scales ranging from the global to the microscopic. Craters or their traces can be found on every solid surface from Mercury to Pluto and surely beyond. As the implications of this plainly observed fact sank in, the idea of impact-driven catastrophes began a slow crawl back to scientific respectability.

Ironically, living on Earth had contributed heavily to the misperception that catastrophes were not important because, here, impact craters are comparatively rare and hard to recognize. Of all the planets, Earth has the most active atmosphere and hydrosphere. Wind, rain, and waves erode its surface quickly. On longer time-scales, plate tectonics and volcanic activity recycle the crust, destroying the old and bringing forth new. Ancient in human terms, most of Earth's surface is incredibly young in geological terms and this induces a profound nearsightedness to which even geologists weren't exempt.

In 1980, the field of planetary science metaphorically tapped earth science on the shoulder and bluntly informed paleontologists that they live and work on a planet. A team headed by Nobel Prize winner Luis Alvarez proposed that the famous great extinction 65 million years ago that ended the Cretaceous Period and wiped out the dinosaurs was caused by the impact of a body, either an asteroid or a comet, that was about 5 to 10 miles (10–15 km) across. As evidence, the Alvarez team pointed to telltale traces of the element iridium found all over the world in sediments from the extinction layer. Iridium occurs only rarely on Earth, but it is abundant in certain types of meteorites.

As might be expected, most earth scientists responded with scorn when the idea was proposed. Though phrased differently, reaction ran along the lines of "Just who do these rocket scientists think they are, telling us we don't know our business?" Tempers at professional meetings and in publications ran high for several years, and the rancor echoes still in certain academic departments. But over the years, extensive field work has turned up abundant evidence, including the probable crater itself, which lies buried under the sediments of Mexico's Yucatán Peninsula. (Some evidence indicates there may also have been more than one impact around the same time.) By now, while there is still much debate about mechanisms — how impacts cause extinctions — few knowledgeable scientists dispute the basic scenario.

The dinosaur-killer impact occurred 65 million years ago, and it would be only human of us to assume that our lives are safe from ever being disturbed by any similar event. This would be a dangerous delusion. The fact is that nothing except luck and the law of averages is preventing a similarly-sized comet or asteroid from striking Earth tomorrow. Indeed, Earth receives large impacts on average every 100 million years or so. If a blow like that of 65 million years ago landed today, it would almost surely wipe out humanity along with much of the life that shares the planet with us. Even a somewhat smaller impact, of the kind that occurs roughly every 20,000 to 200,000 years, would cause tremendous harm to human civilization and the ecosystems we depend upon. Scientists distinguish between "extinction level events" (such as the dinosaur-destroying impact) and "civilization-killers," which could blast mankind back to the Dark Ages or further. But even the lesser is bad enough.

The danger is not that Earth would be knocked on its side or thrown from its orbit around the Sun, as supermarket tabloids are prone to claim. No impact by a comet or asteroid is likely to do that — they are too small by far and Earth too large. Instead, the threat of a major impact lies in its ability to inflict more damage on Earth's *atmosphere* than it can absorb and neutralize before most living things, including us, are dead.

When the 20-plus fragments of Comet Shoemaker-Levy 9 slammed into Jupiter in July 1994, it gave everyone on Earth a first-class demonstration of the random violence that nature can wreak with a comet. Television brought the view to the world with startling sights captured by the Hubble Space Telescope and other observatories, including the Galileo spacecraft then approaching Jupiter. Day after day for more than a week, huge "bruises" of dark dust — each the size of Earth — blossomed in Jupiter's southern latitudes, as the freight train of comet pieces plunged into its doom. The impact sites were visible to anyone with even a small telescope, and the splotches took months to fade from detectability, despite hurricane-velocity winds that rip Jupiter's cloud-belts into streamers.

Some lessons of the great comet crash went beyond the obvious. For the scientists who study impacts and their effects, one of the revelations coming out of the event was the importance of ejecta thrown out of the primary impact. When each comet fragment screamed down into Jupiter and exploded, it shot a geyser of vaporized debris up and away from the impact site. This material sailed off on ballistic paths and re-entered the jovian atmosphere at high speed at some distance away. This caused a secondary wave of impact heating that carried the blast effects to parts of Jupiter that were not directly affected by the main impacts.

Something similar must have happened 65 million years ago. Traveling toward the northeast, the object struck a half-mile-deep ocean covering what is now Mexico's Yucatán region. The impact blasted a crater in the sea floor that was at least 120 miles (200 km) across and 20 miles (30 km) deep. The crater was deep enough to reach through the

End of an era. Countless times in Earth's history, our planet has been struck by comets and asteroids. Most impacts are small and do little more than local damage, but big impacts, fortunately rare, may cause global mass extinctions and abruptly change the course of evolution. We benefited from one such impact-driven extinction 65 million years ago, when it wiped out dinosaurian life and let mammals diversify and thrive. (Don Davis, NASA)

crust to the soft, hot rocks of the upper mantle. Earthquakes larger than anything in human experience shook the ground for thousands of miles. The impact sent out enormous sea waves that ripped up ocean bottom sediments as far away as south-central Texas and threw a thick layer of shattered and molten rock over central Mexico and Haiti. The entire Caribbean Sea may have been emptied of water — briefly.

Some debris simply shot straight up into space and never came back. Other debris, flung on lower and slower trajectories, re-entered the atmosphere over more distant parts of the globe, causing such a pulse of heat that it ignited whole forests and left a telltale layer of soot that can be traced over much of the Earth. (When you see the bright streak of a meteor burning up on a dark night, you won't feel the slightest flicker of heat from its light, and even a rich meteor shower sheds no tangible warmth. These particles, however, are barely the size of a grain of sand or perhaps a pea. The picture changes drastically when the "shooting stars" are each many yards across and the sky is completely filled with them.)

Tons of fine rock particles, chemical aerosols, and smoke from the widespread fires were carried high into the stratosphere where they formed a global shroud that shut off sunlight for months or years. Stygian darkness and bitter cold prevailed for months over most of the Earth, closing down plant photosynthesis and killing much vegetation. This doomed the animals that fed on plants, and their deaths in turn spelled the end for animals that fed on herbivores. Since all life depends on photosynthesis, directly or not, the disaster's lethal effects went up and down the food chain, striking most heavily at animals who were large or unable to hibernate, or who could not eat carrion and forage in the detritus of a ruined world.

Skies cleared only when snow and rain slowly washed the particles to the ground. But the precipitation was no friend to surviving creatures on the surface — even the rains were hostile, made strongly acidic by atmospheric chemical reactions which also destroyed the ozone layer. The acid rain probably also turned the upper layer of the oceans toxic.

As the final act in this cavalcade of disasters, the sunlight let in by clearing skies swiftly boosted temperatures to life-threatening levels, through an enhanced greenhouse effect. This was caused by abundant carbon dioxide in the atmosphere, which came from the vaporized rocks at the site of the impact. By a stroke of incredible bad luck, these rocks combined layers of limestone and anhydrite — calcium carbonate and calcium sulfate. Decomposed by the heat of the blast, they yielded copious amounts of sulfur aerosol particles (producing acid rain) and CO_2 (producing the greenhouse).

After thousands of years of chaos, destruction, and death, the agony subsided. Skies turned blue again, vegetation bloomed and spread, and cicadas droned in the days of summer. Sunlight once more sparkled on the waters. It seemed like everything had returned to normal. Yet it hadn't — and never will. The dinosaurs and much of their world were gone forever. New plants and animals had evolved to occupy empty places in the world's cycle of eat-and-be-eaten. We are the beneficiaries of that great extinction, since the mammals we descended from had only limited opportunities when the giant lizards ruled the world. We owe our existence to a terrible impact that changed the course of life on our planet, but we should never for a moment think it can't happen again.

If comets can hit Jupiter, they can just as easily hit Earth. And as it happened, Earth *was* hit by a very small asteroid or comet on June 30, 1908 — although luckily the object was only about 200 feet (60 meters) across and the impact point lay near the Stony Tunguska River in an almost deserted part of the Siberian taiga.

A lot of nonsense has been written about the Tunguska event, with various accounts proposing that it was caused by an exploding alien spaceship, the impact of a small black hole, or the arrival of a piece of anti-matter. The craziest notions have sprouted largely because the impacting object left few physical remains. However, in recent years, studies that combine computer simulations with detailed fieldwork have drawn a plausible picture of what must have happened that day.

The Tunguska impactor descended at about 60° from the vertical on a path traveling northwest at a dozen miles per second. As the shock wave of air in front of the swiftly moving object rapidly grew denser, the object began to break up. Shattering completely in milliseconds, the impactor abruptly halted as if it had hit a brick wall. (If this sounds improbable, recall the last belly-flop you made into a swimming pool. Under certain conditions, normally yielding substances — air, water — can behave like concrete.) Brought instantly to a stop, and with so much kinetic energy to disperse, the object exploded in the air at an altitude of about 4 or 5 miles (6–8 km), smashing flat more than 2,000 square miles of forest.

Because the area was remote and other national affairs in Russia took precedence, the first scientific survey did not reach the site until 1927. When members of the expedition eventually arrived, they found no visible crater but millions of blown-down, charred trees. Those at ground zero (the point at the center of the devastation) were left standing but had their branches stripped off, as if a blast wave had come from directly above. For miles all around, fallen trees lined up pointing radially away from ground zero. From the charring on the tree-trunks, it seemed like a tremendous flash of heat had set the trees alight and then a shock wave immediately afterward had snuffed the fires like candles on a birthday cake.

No one knows how many people were in the area of destruction at the time of impact (about 7 in the morning, local time). Interviews with surviving witnesses indicated that the loss of life was probably small, thanks to the sparse population. The eye-witness reports, however, are dramatic. At a distance of 45 miles (60 km) from ground zero, a man on the porch of a trading post was blown over by the shock and his shirt became so hot it caught fire. Several reindeer herders and their families who had the bad luck to be closer to ground zero were thrown into the air and knocked unconscious. One was hurled about 40 feet into a tree and killed. Most herders in the immediate area lost their reindeer, and the remains of many animals were found in various states of incineration. Farther afield, across Europe and much of the world, seismographs, barographs, and magnetic field instruments detected the impact and the passage of its blast wave. And even city dwellers in western Europe noticed that the night of June 30 to July 1 was light enough to read a newspaper or a book outdoors at midnight.

Researchers today have concluded that the impacting object was probably a carbonaceous-chondrite-type asteroid or perhaps an extinct comet nucleus. (Eugene Shoemaker, one of the co-discoverers of Comet Shoemaker-Levy 9, once called extinct comet nuclei the "stealth

One of Earth's freshest impact craters is Meteor Crater in northern Arizona, 1,300 yards across. Created 50,000 years ago by the impact of a house-sized piece of nickel-iron, this structure is a pinprick compared with the much larger craters that still mar the faces of planets and moons throughout the solar system. On Earth, erosion quickly erases the scars of impacts, leading us to assume — incorrectly — that we are safe from such catastrophes today. (David Roddy; US Geological Survey)

bombers" of the solar system, noting they are dark and hard-to-find and often travel on Earth-crossing orbits.) If the Tunguska object was a comet fragment, one possible source is Comet Encke. In any case, the object was clearly not made of nickel-iron, like the meteorite which blasted out Meteor Crater in northern Arizona 50,000 years ago. The Tunguska impactor left no crater because the incoming body was destroyed before reaching the ground. While this object was small in cosmic terms, the blast from the impact was rather impressive all the same: it about equalled the detonation of a 15 megaton nuclear bomb, except for the latter's radioactivity.

The Tunguska airburst occurred over a remote forest. Had it struck a populated area, the loss of life would have been beyond calculation, and no one today would need to be convinced that comets and asteroids pose a threat.

Unfortunately, humanity seems to have a persistent blind spot in this regard. The 1994 impact of Comet Shoemaker-Levy 9 with Jupiter breathed new life into a NASA study called *The Spaceguard Survey*, chaired by David Morrison of NASA's Ames Research Center in Mountain View, California. Carried out in response to a US congressional directive, *The Spaceguard Survey* was completed in 1992 but largely ignored after publication. As conceived in the study, Spaceguard is a plan for a worldwide network of half-a-dozen automated telescopes, each about 2 to 3 meters (80 to 120 inches) in aperture, dedicated to finding comets and asteroids that might strike Earth. The study estimated that within 25 years of operation such a network could find more than 90% of the Earth-approaching objects that pose a serious threat to us.

Regarding the costs of the survey, the Spaceguard report said, "The initial cost to build six 2.5-m telescopes and to establish a center for program coordination is estimated to be about $50 million (FY93 dollars), with additional operating expenses for the network of about $10 million per year... Over the first decade of operation the survey would require appropriations approaching $100 million, perhaps half of which could be provided by the United States and half by international partners."

When considering threats to Earth, scientists use a catchall term for potential impactors — they call them Near Earth Objects, or NEOs for short. The chosen term is deliberately non-specific. Searches for NEOs do not discriminate between comets and asteroids because the same techniques find both kinds of object, and potentially dangerous impactors can come from anywhere in the sky irrespective of type. Besides, the comet/asteroid distinction doesn't matter too much when an object strikes at a velocity of a dozen miles per second or more. At that stage, whether it's made of ice or rock is a secondary issue.

Which raises a point. Many people become confused by the distinctions planetary scientists draw between meteors, meteorites, meteoroids, asteroids, minor planets, comets, and the like. Forget the names for a moment and remember just this: objects that can strike Earth range in size from microscopic dust to solid bodies dozens of miles across. Impacts from the tiny end of the range produce pretty light shows like meteor showers, but an impact from the small, medium, or large end of the range could destroy a city, wipe out human civilization, or kill most of the life on our planet. Spaceguard is an attempt to inventory the small, medium and large objects.

Spaceguard's goal is to discover every object one kilometer (half a mile) or bigger in size, track its orbit, and see if it poses a collision danger to Earth. The threshold of 1 kilometer was chosen because this is believed to be the smallest object whose impact could do enough damage to the world to threaten the survival of civilization. If Spaceguard were to find any such object on a collision course, then the foreknowledge might give us enough time to prevent the impact or at least take precautions for minimizing the damage.

The question of what to do if an object were found on a collision course was the focus of a second study authorized at the same time as the Spaceguard survey. Unfortunately, its results were much less clear-cut. The task was carried out by scientists working in the military and defense communities, with some involvement by civilian scientists. However, the interception plans quickly became entangled in secretiveness (perhaps inevitably) and they ended up giving the impression that they were mainly a way for various weapons programs to stay in business after the Cold War ended. Since then, military programs (both search and interception) have pursued a track that largely parallels civilian programs, but mostly doesn't mingle with them.

For many people who hear of Spaceguard, the plan strikes them as unbelievable or ridiculously farfetched. When published, it earned among politicians what one of them described as a "high giggle factor." The Jupiter impacts of July 1994 erased the smiles, however, and that same month, NASA, under Congressional prodding, asked a group led by Eugene Shoemaker of Lowell Observatory to update the Spaceguard plan in light of new technical developments. The Shoemaker group found that a Spaceguard survey could be completed in 10 years at a total cost of less than $50 million — somewhat more favorable figures than the original estimates. (To put these numbers in perspective, in 1998 the movie *Deep Impact* earned box-office receipts of almost $130 million in its first six weeks of release, and the scientifically much inferior *Armageddon* raked in $149 million in its first four weeks.)

Yet, even with new technology, Spaceguard still seemed very pricy, at least to those who would have to lobby for it in Congress. So instead of pursuing the broad political backing that a formal start on Spaceguard would require, NASA initially settled for contributing a small amount of money (about $1 million a year) toward existing search projects, mainly as a way to test techniques.

In 1998, however, following a shift of focus, NASA bumped up NEO research in priority. In 1999 it contributed $3.5 million toward comet and asteroid searches at a number of observatories. Also in 1999, the US Congress authorized additional search funding, up to $10.5 million a year for three years, but this faces many budget-process hurdles before being enacted.

Among these are the Spacewatch project at Kitt Peak National Observatory headed by Tom Gehrels of the University of Arizona, and the Lowell Observatory Near Earth Object Search (LONEOS) at Flagstaff, Arizona, run by Ted Bowell. Also in Arizona is the Bigelow Sky Survey on Catalina Mountain, run by Stephen Larson of the University of Arizona. Another search project is the Near Earth Asteroid Tracking (NEAT) program, run in collaboration with the US Air Force and the Lawrence Livermore National Laboratory using an Air Force telescope at Haleakala on Maui, Hawaii. The director is Eleanor Helin of Caltech's Jet Propulsion Laboratory. A fifth program, which is proving to be the most

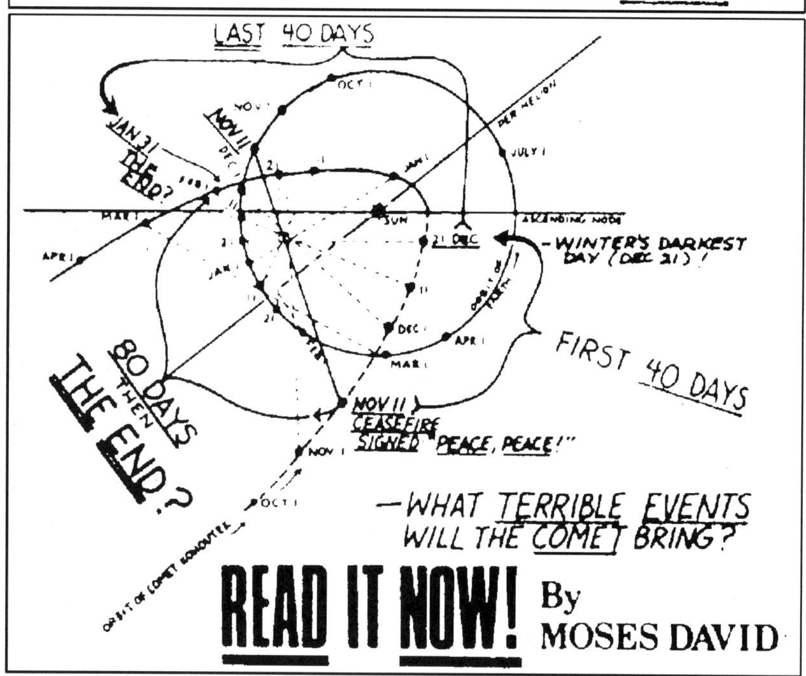

Fire and brimstone! During the period when it was still being hailed as the "comet of the century," poor old Comet Kohoutek was dragged into the portents-and-omens game. (From a flyer collected in 1973 by the author)

productive of all, is the LINEAR search, run by MIT's Lincoln Laboratory near Socorro, New Mexico, under the direction of Grant Stokes. In addition, NEO discovery and follow-up efforts are underway at the University of Victoria (Canada), the Observatoire de la Côte d'Azur (France), and in China. Most of the search programs use telescopes of around 1-meter (40-inch) aperture, and Spacewatch is soon to start using a 1.8-meter telescope. Modern electronic sensors and automated detection software identify NEOs moving against the background star fields.

To coordinate NEO search efforts, in July 1998 NASA created a new program office at the Jet Propulsion Laboratory under the direction of Donald Yeomans. The program goal is essentially the same as Spaceguard's: locate at least 90% of the estimated 2,000 asteroids and comets that approach Earth and which are 1 kilometer in diameter or bigger, and do this by 2010. (Less than 300 NEOs are known at present.) The program is aiming to achieve the main goal of Spaceguard at a lower cost by upgrading existing telescopes and detectors. NASA also hopes to work with the US Air Force to use more of its 1-meter telescopes that were originally built to track satellites from the ground.

The problem with the searches up to now is that, while they have proved that the concept indeed works, they have been chronically understaffed, underequipped, and underfunded. And discouraging political hurdles remain. For example, a productive Australian NEO search was shut down in 1996 for budget reasons, despite international protests from scientists. Even now, NASA contributes only a quarter of the funding for the University of Arizona's Spacewatch program; most of the rest comes from private donations. Moreover, many scientists are concerned that the new "mini-Spaceguard" won't reach out to faint enough magnitudes to pick up all the hazardous objects, nor will it cover the sky as thoroughly as needed. The result of such skimping is that the astronomical fishing net is woven very coarsely and countless NEOs undoubtedly escape detection through its gaps. Statistics show the discovery rate is only about 20% of what it should be to meet Spaceguard goals. And, as David Morrison puts it, on a typical night you'll find more people behind the counter at a McDonald's restaurant than there are looking for objects that could put an end to humankind.

The risk is real. Incomplete or not, current surveys turn up objects that go whizzing by too close for comfort. In 1989, one passed at twice the Moon's distance, and in January 1991, an object zipped by at less than half the distance to the Moon. In May 1993 another came by a little closer, and in March 1994 so did another. In December 1994, one missed us by less than a third the Moon's distance. Nearly all of these were found only a couple of days before they made their closest approach to Earth. Moreover, because the surveys are limited, astronomers are painfully aware that searches are catching only a fraction of these near-misses. Estimates indicate that about one such object flies past us every week, noticed or not.

Well — a cynic might say — so far, so good. All these NEOs have missed, and all were relatively small (a few tens of yards in diameter). But if any of them had had Earth's name on them, the area at ground zero would have received too little warning to do anything — much less time, for example, than a coastal region gets when a major hurricane or cyclone bears down on it.

Warning time is everything. The 18 days depicted in *Armageddon* is a grotesque joke, and even the two years shown in *Deep Impact* is in reality too short for building a spaceship to fly an intercept mission such as the movie portrays. (A decade is more like it.) Actually, the best that the fictional world of *Deep Impact* could have done, after identifying ground zero, is to evacuate the affected regions. However, the idea of bunker-like "arks," in which at least part of the population could be sheltered, was realistic. Something of the kind would be sensible to prepare if humanity finds itself forced to ride out an impact and the aftermath in which billions of people will probably die.

A non-fiction impact scare occurred in March 1998. On this occasion astronomers had the name and number of the rock, and could even point to a date on the calendar. That day, it transpired, was just 30 years off, perhaps too soon to do much about deflecting the asteroid if it really proved to be on a collision course. Moreover, as a dress-rehersal of a real impactor discovery, the event revealed a couple of flaws in the system.

Here's what happened: In the course of analyzing observations of a small asteroid discovered by Spacewatch — 1997 XF11 — Brian Marsden found the object had an orbit that might bring it within 26,000 miles (42,000 km) of Earth's surface on October 26, 2028. It might also pass by much farther away. Marsden, who heads the Minor Planet Center at the Harvard-Smithsonian Center for Astrophysics, could see that the numbers indicated that XF11 would probably miss in all cases. But it looked like it was coming pretty close and a collision, he felt, was not entirely out of the question. Since estimates placed the asteroid's diameter in the range of a quarter-mile to 2.5 miles (0.5 to 4 km), any impact would likely have extremely serious consequences no matter where it struck.

On March 11 Marsden put out an e-mail *IAU Circular* (#6837) about the close passage in 2028 and asked for more observations of XF11. Couched in technical language, the *Circular* was not something that would alarm the general public. But anybody with astronomy experience, such as science reporters, could grasp the implications immediately. (I certainly did.) In addition, Marsden prepared an information sheet for the media and general public that spelled things out unsensationally but a lot more clearly. It also provided background on potentially hazardous objects. He sent this document to an e-mail distribution list for press releases run by the American Astronomical Society. Reporters got hold of this on the afternoon of March 11 and by the next morning XF11 made headlines everywhere in the world.

Another day later, however, the scare was over. What happened? In the interval, other scientists involved with comets and asteroids had made their own orbit calculations using the original data and a few pre-discovery observations that rummaging in the archives quickly turned up. These calculations revised the most-probable miss distance to 600,000 miles (960,000 km) and confirmed that XF11 was *not* going to hit us in 2028. After word of this spread to the media via more e-mailed press releases, the general public breathed a sigh of relief. Headlines now proclaimed "Doomsday Cancelled!"

Well, postponed at least. This particular asteroid has an orbit that brings it near Earth's orbit every few years, and it will make many more close passages by our planet in the decades to come. Calculations indicate that we are indeed probably safe from 1997 XF11 for the

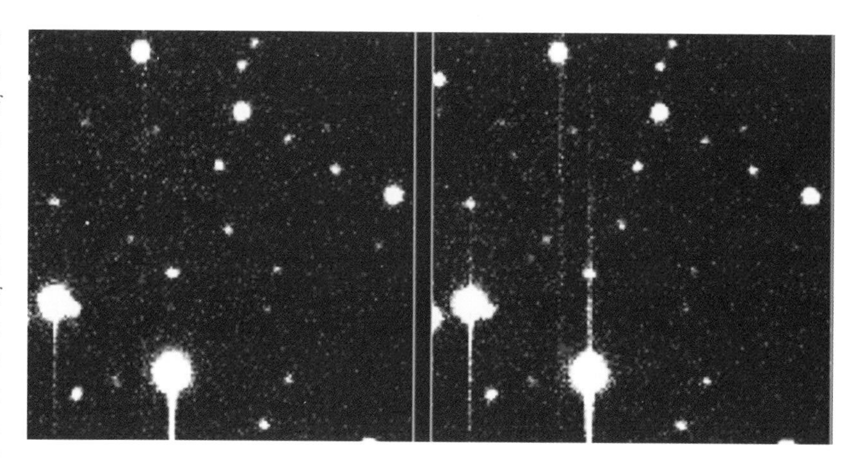

Doomsday cancelled? Discovered by the Spacewatch team at the University of Arizona, the one-mile-wide asteroid 1997 XF11 (moving dot, center) will pass close to Earth in 2028. The goal of Spacewatch and other comet and asteroid discovery programs is to see if any object out there poses a risk of collision. If so, the sooner we learn of it, the better our chances for diverting its course. (University of Washington; Astrophysical Research Consortium/Apache Point Observatory)

next century at least, but longer projections are increasingly uncertain because XF11's orbit cannot be determined over long periods. The immediate concern may be over, but a very small risk still remains.

Back in the world of planetary science, however, 1997 XF11 generated its own kind of impact. The announcement and retraction a day later created the public impression that astronomers just can't get their numbers right. In fact, correction and revision are always part of science, though they aren't usually carried out amid such a roar of publicity. Many planetary scientists felt that Marsden could have handled the affair much better. For example, they said he should have privately queried NEO observers for additional observations, instead of using the semipublic *IAU Circulars*, which are subscribed to by many besides professional scientists. Their biggest complaint, however, was that he had not let other orbit calculators using different software and analytical routines double-check the miss distances and collision probabilities before "going public." After all, they argued, what harm could be done by a 24- or 48-hour delay regarding an object that won't come close for over 30 years?

One main result of the affair is that NASA and other organizations involved set up a process whereby scientists outside the Minor Planet Center will verify any comets or asteroids on suspect orbits before similar announcements are made in the future.

The system got a real-life test a year later, when another asteroid was found with an orbit that conceivably could strike Earth. Asteroid 1999 AN10 was discovered by LINEAR, and appears to be about a mile (1.6 km) in diameter, big enough to be of concern. Orbital calculations by a group headed by Andrea Milani at the University of Pisa in Italy found that the asteroid makes many close approaches to Earth. The asteroid, they calculated, poses no threat for the next 40 years, although an approach in August 2039 will come close enough to pay attention to. But even in this case they computed that the probability of impact was about one chance in a billion. Posting the analysis on their web site, they asked other scientists to check their work but did not issue any general press release about the asteroid. The reviews, which were done by orbital dynamics scientists around the world, agreed with the Pisa group's conclusions.

However, an anthropologist in England, Benny Peiser, was bothered by the lack of public notice regarding the news that AN10 posed an impact threat. After reading the web-posted paper, Peiser circulated an e-mail message on a listserver that he runs which is concerned with the effects of catastrophic events on Earth, including impacts. Peiser said he was disturbed by the quiet treatment AN10 got compared with 1997 XF11, and he wondered it it didn't reflect an overreaction to the embarassment that the XF11 affair caused planetary scientists.

Stung by the charge of secrecy, scientists overwhelmingly responded on Peiser's listserver that what they were doing was the best, indeed the only, way to deal with such findings: thorough peer-review before taking the findings to the public. The discussion on the listserver reached some in the general media who took up the story. But this time, however, the uproar was much less than in the case of XF11, perhaps because the media felt thay'd been burned once — and perhaps also because many science reporters basically agreed with the scientists' position.

In an attempt to place impact probabilities on an easily comprehended scale for alerting the public, planetary scientists attending a 1999 conference in Italy developed the "Torino Scale," named for the city where they met. The Torino Scale runs from 0 (no chance of an impact in the foreseeable future) to 10 (a major impact will occur for certain). Applying the scale to the cases mentioned above, both rate as zero. Scientists are hoping that the scale will prevent future overreactions on the part of the media, while helping people and governments assess realistically whatever threat any particular object poses.

For many decades, the United States and the Soviet Union (now Russia) have had satellites in orbit looking down for the hot bright trail of a missile being launched or the distinctive pulse of a nuclear explosion. Over the years, these satellites have patiently monitored the planet with instruments designed for military purposes. By their nature, however, these instuments are also well-adapted to spot the infrared trail that a large meteorite leaves in the atmosphere or the flash of an impact. Between 1975 and 1992, spaceborne detectors belonging to the US logged 136 impacts of meteoroid material, each with an energy of 1 kiloton or more. More impacts than these certainly occurred, but because the system was oriented toward missile detection, the data for clear-cut cases of natural events were not always logged and preserved. All the logged impacts were airbursts like Tunguska, but smaller in size. Surprisingly, on average about 80 impacts of kiloton-size occur each year. (This estimate is based on the number of events seen by the satellites' detectors, those missed by its instruments, and those noticed but disregarded because they were obviously natural.)

During the Cold War there was no way for this data to reach civilian scientists because the instruments, their capabilities, and even the existence of the satellites were classified top-secret. (The satellites' full capabilites remain classified still.) However, the gradual emergence into the open literature of missile-monitoring data has shown that impacts occur remarkably often, essentially on a weekly basis.

So why aren't people making more fuss over them? For the most part impact explosions pass unnoticed by people on the ground because they occur over empty reaches of ocean or sparsely inhabited parts of the world. Also, because most impactors are stony (or icy, in the case of comets) they are physically weak. This means that aerodynamic forces break them up and disperse them at high altitudes. Small cometary debris probably won't make it below about 20 miles (30 km), while tougher stony objects could survive down to 15 miles (20 km). Real damage begins with objects that are big enough or sturdy enough to reach 10 miles' (15 km) altitude or less. This calls for a rocky object bigger than about 150 feet (50 meters) across, or a cometary one at least 300 feet (100 meters) in diameter.

Besides exploding at high altitude, simple non-recognition probably also plays a large role in the general public ignorance. It's likely, for example, that some of the sonic booms that people automatically blame on military jets are actually shock waves from incoming objects. Recall too those medieval prodigies of nature which included thunder booming out of clear skies. Wouldn't it be darkly funny if some of those actually *were* comet impacts?

Non-recognition can take numerous forms. Work by geologist Ted Bryant of Australia's University of Wollongong has found deposits of ocean-bottom debris on the eastern and western coasts of Australia that were left by tsunamis (tidal waves). The deposits show that in places the giant sea waves crested at heights of more than 200 feet (60 meters) and washed heavy boulders about 20 miles (30 km) inland. Radiocarbon dating of the debris indicates the most recent of the tsunamis occurred a little more than 200 years ago, shortly before the European settlement of Australia.

From geological details in the deposits Bryant discounts the known causes of tsunamis such as undersea eruptions, landslides, and earthquakes as sources for the giant waves. He suggests instead that they could have come from the impact of a comet or, more likely, several fragments of one comet, since the coastlines don't face a common ocean. If true, it provides an explanation for an Aboriginal legend from the southeastern coast that speaks of a white wave falling from the sky. Up until now this has been taken as a metaphorical description of the arrival of European settlers, but it could also be a factual account of a giant impact-tsunami engulfing the coastline. It's possible that many impact events lie encoded in such tales and local myths.

Finally, there are questions of scale and frequency. Statistics show that an impact producing a blast the size of the Hiroshima bomb occurs about once a year, and a megaton event can be expected roughly once a century. It seems hardly believeable that such events would pass unnoticed, but remember that most will be high-altitude airbursts occurring over the open oceans where few witnesses will see them. Moreover, the majority of satellite-recorded impacts are quite small, involving objects

perhaps only a yard or two across. This means the whole phenomenon is just not very dramatic unless it happens close by.

For the record, the largest satellite-recorded impact came on February 1, 1994, near Kusaie Island in the western Pacific Ocean about 1,000 miles (1,600 km) northeast of Papua New Guinea. As it descended at about 45° and traveled northwest at 15 miles per second (24 km/s), the object broke up in two stages. A small explosion occurred at an altitude of 21 miles (34 km) and a final much bigger bang came at 13 miles (21 km). The flash was nearly as bright as the Sun. Using the brightness of the flash as detected by satellites plus statements from a pair of badly startled native fishermen who happened to be in the area, scientists estimated that the original object was the size of a typical house and stony in composition. The blast energy was in the range of 50 to 100 kilotons, or 3 to 7 times stronger than the nuclear explosion that destroyed Hiroshima. But as it had only about 1/200th the power of the Tunguska impact and was a high-altitude airburst, there was no damage to the surface and luckily (since it was over open ocean) no tsunami either.

All right — bottom line — what's the risk? How likely is it that something the size of an office building or larger will slam down over New York City, say? It's hard to place the odds in any kind of framework comparable to our everyday experience. Unlike more routine threats that people face with relative equanimity — car accidents, airplane crashes, food poisoning, slipping and falling in the bathtub — the possibility of a big comet or asteroid impact combines a very unlikely event (taken over the average lifetime of people alive now) with a stupendously high level of damage should it occur. When you multiply a very small number (the odds) against an extremely big one (the damage), the result is not a negligibly small number and it should be taken seriously. It turns out to be comparable to the chances of someone dying in a plane crash: about 1 in 20,000. Just for comparison, the chances of a U.S. citizen dying in a car crash are vastly worse: 1 in 100 (assuming a 50-year driving career). For losing one's life in a homicide, it's 1 in 300. The odds of being killed by a flood are about 1 in 30,000 and of being a victim of a tornado is 1 in 60,000. For death by botulism, it's 1 in 3,000,000. If you relish life's ironies, it's bleakly amusing to watch society spend bags of money trying to reduce the threat of ordinary disasters (to coin a phrase), while

The unpredictable arrival of Great Comets teaches us that we are living inside a long-running story — the formation of the solar system — that has yet to reach an end. (Comet Hale- Bopp; Ben Gendre; 5.5-inch Schmidt camera, 5 minutes on Kodak Royal Gold 100; April 5, 1997)

turning its back on the one extraordinary disaster that's comparably dangerous. There's the giggle factor in action.

NEO researchers are careful to point out that Spaceguard does not warrant an all-out approach like the Manhattan Project that built the atomic bomb during the Second World War. And certainly Spaceguard should not be funded by taking money from other efforts to reduce risks in everyday life, such as public health or automobile and airplane safety. However, given Spaceguard's relatively small costs and what's at risk, it would be foolish not to pursue it fully. We'd feel damn silly if Spaceguard were carried out half-heartedly or even delayed by several decades — only to have it uncover a comet or asteroid headed our way with an impact date by then only a few years off. The dinosaurs had no knowledge of this kind of threat to their existence, nor any way to deal with it. We do.

In any case, human beings should harbor no illusions about the risk of impacts. Simply look at the Moon some night with a small telescope. For every lunar crater you see, Earth once sported about 20 craters, thanks to its larger size and stronger gravity. Some craters came from comet impacts, others from asteroids. Impacting comets delivered part of our oceans, and deposited water on the Moon where it now resides as ice in the soil at the poles. But unlike the airless, dead Moon, terrestrial erosion and the relentless reworking of crust by plate tectonics have erased the scars of most impacts here. We look around and because we don't find a crater in our own backyard, we're lulled into thinking the solar system's firing squad has run out of ammunition.

As the heavens parade overhead every clear night, the spectacle they present is broadly familiar. Stars form their age-old patterns and planets slowly change position as weeks roll past. The scene is one we've witnessed before on countless nights. Not so with comets — they appear too rarely to become truly familiar. Their scarcity confers a strangeness that adds to the wonder and delight we feel upon spotting them.

Literally older than the hills, comets take us back to our origins. Primitive objects left from the dawn of the solar system, comets remind us that we live within a vast natural story that began before there even was an Earth, let alone life of any kind. And this story will still be unfolding in ages to come, when the species that descend from us are themselves fossils eroding out of ancient rock. Even then comets will be trailing their gossamer sails through the night.

Out for a walk one afternoon in late autumn years ago, I passed a Clouded Sulfur butterfly. It was hopping from one thin salt-cedar leaf to another, flexing bright yellow wings edged in black as it tried to draw warmth from a westering Sun. On my return half an hour later, I saw the leading edge of a winter storm brushing across the mountains that lay some 20 miles to windward. The storm was dropping heavy grey trails of snow as it pushed across the peaks. In the storm's approach, I saw the end of that year's long New Mexican autumn. I also knew I was seeing the Clouded Sulfur's impending death. It would lose its life within the hour to sheets of freezing rain that were coming on winds it couldn't escape. There was nothing, really, the butterfly could do to save itself — or that I could do for it. But on returning to the house I brought in several armloads of wood for the stove.

I sometimes wonder how humanity might look to a Great Comet flying past on its long orbital journey. Our time scale and the comet's are even more mismatched than mine and the butterfly's. Human perceptions dwell relentlessly in the here-and-now, which is both a blessing and a curse. Science tells us over and over that all life is an experiment-in-progress, with no particular outcome preordained. Whatever can said about other animals, we have a self-consciousness that bestows the ability to envision futures. I hope that visits from Great Comets from time to time can help us to see our planetary home and its surroundings clearly and without illusions — and encourage us to take all the steps within our power to keep it habitable.

8 Staying Current With Comets

No one can predict when and where the next Great Comet will appear, because we have no way of knowing what's heading our way from the Oort Cloud or Kuiper Belt. The next Great Comet may be discovered tomorrow — or it may be 10 years or more from now. In the meantime, many beautiful comets will come and go, and it would be a shame not to enjoy them to the fullest. The resources listed below will keep you up to date on what's happening.

Comet web sites

Make it a habit to check the following web sites frequently, as comet discoveries appear much faster than information sources such as magazines can get news of them into print. If you are not online at home, try your local library. Many now offer public-access terminals you can use. Ask at the Reference or Information Desk.

For simplicity's sake, the web site addresses given below omit the usual browser command "http://". Bear in mind that web addresses change from time to time, although the organizations behind these sites are good about keeping them stable or at least providing forwarding links. (If all else fails, try using keywords with a search engine like Alta Vista or Yahoo.) These addresses would make a good starter set of bookmarks for any Internet-surfing comet fan.

New comet discoveries are published in the *International Astronomical Union Circulars* by the Central Bureau for Astronomical Telegrams at the Smithsonian Astrophysical Observatory in Cambridge, Massachusetts. (Despite the Bureau's name, the *IAU Circulars* arrive by e-mail.) For more information on how to receive them, go to **cfa-www.harvard.edu/cfa/ps/cbat.html**, or send an e-mail note to **iausubs@cfa.harvard.edu** asking for details.

The Minor Planet Center at the Smithsonian Astrophysical Observatory maintains a web page for staying on top of the growing numbers of Kuiper Belt objects (the so-called transneptunian objects or TNOs) and

Centaurs. Take a look at: **cfa-www.harvard.edu/iau/lists/TNOs.html** and **cfa-www.harvard.edu/iau/lists/Centaurs.html**.

For information regarding the searches for comets (and asteroids) that might hit Earth, check out a web site at NASA's Ames Research Center, **impact.arc.nasa.gov/**. In the "Reports" section, you'll find the text of *The Spaceguard Survey* report (and others), and in "Related Materials" are links taking you to the home pages of the Spacewatch and other search programs and much else. (The Torino Scale for impact hazards is detailed fully here, too.) The Minor Planet Center has a useful page for Near Earth Objects (NEOs) at **cfa-www. harvard. edu/iau/NEO/TheNEOPage.html**. It also keeps an up-to-date list of potentially hazardous objects at **cfa-www.harvard.edu/ iau/lists/ Dangerous.html** and a page giving the dates of upcoming close approaches to Earth at **cfa-www.harvard.edu/ iau/lists/ PHACloseApp. html**.

The *International Comet Quarterly* has a web site run by Daniel Green of the Central Bureau of Astronomical Telegrams. Its URL is **cfa-www.harvard.edu/iau/icq/whatisICQ.html**. The *ICQ* is a scientific journal devoted to the observation and study of comets. It links amateur astronomers with professionals for exchanging observations and news.

At the Jet Propulsion Laboratory (JPL), Charles Morris operates the Comet Observation Home Page at **encke.jpl.nasa.gov/**. This site records observations (visual and photographic) for all comets currently or recently visible, whether great or small. It includes links to other good comet sites too.

Also at JPL, Ron Baalke runs two web sites with lots of news and images on Comets Hale-Bopp and Hyakutake. The Hale-Bopp page is at **www.jpl.nasa.gov/comet/** and Hyakutake one is at **www.jpl.nasa.gov/comet/hyakutake/**. Both have links to many other sites.

The Space Telescope Science Institute, which operates the Hubble Space Telescope (HST), will report on any comets HST observes. Check out their public information page, **oposite.stsci.edu/pubinfo/pictures.html**.

Sky & Telescope magazine operates a comet home page at **www.skypub.com/sights/comets/comets.html**. The magazine has many long-standing connections with comet discoverers and observers, and its web site reflects this depth of involvement.

Astronomy magazine also covers comets. Go to **www.kalmbach.com/astro/astronomy.html**. and use "comet" as a keyword with the site's search engine, or check the sitemap.

Space mission web sites

Numerous comet missions are in development by NASA and the European Space Agency. To stay abreast of changes and news, check their web pages often. Some projects also offer listservers to keep you automatically appraised of developments by e-mail.

The European Space Agency's Rosetta mission is being built and operated by space labs in several countries, so tracking down information can be a somewhat roundabout process. However, the place to start is **sci.esa.int/rosetta/**. (There is also a useful summary of the mission at **stardust.jpl.nasa.gov/comets/rosetta.html**, although since this site is not run by the Rosetta mission itself, it may not be as up-to-date.)

NASA's Stardust mission is operated by the Jet Propulsion Laboratory; the project's home page is at **stardust.jpl.nasa.gov/top.html**. The Comet Nucleus Tour (Contour) is at **www.contour2002.org.**

Information about JPL's Deep Space 1 and Deep Impact missions can be found at **nmp.jpl.nasa.gov/ds1/** (for DS1) and at **www.ss.astro.umd.edu/deepimpact/** (for Deep Impact).

The Pluto-Kuiper Express has a web site at **www.jpl.nasa.gov/ice_fire//pkexprss.htm**.

The Ulysses Comet Watch web page is at **lasp.colorado.edu/ucw/ucw.html**. Although the project has ceased, the web site contains images of Hale-Bopp that focus on the ion tail. Other comets may be added in the future.

While not a comet mission, the Solar and Heliospheric Observatory (SOHO) caught some 60 comets as they collided with the Sun or passed very close to it. Reports of SOHO comets are posted at the page run by Douglas Biesecker at **sungrazer.nascom.nasa.gov/comets.html**. The comets were spotted in data from the LASCO instrument; its home page is **lasco-www.nrl.navy.mil/**. The home page for the SOHO mission is at **sohowww.nascom.nasa.gov/**.

Chapter 5 described NASA's Discovery program, its ongoing series of low-cost solar system missions. Stardust, Contour, and Deep Impact are all Discovery missions, and other comet-related probes are likely to appear in future rounds of mission proposals. To stay up on developments, from time to time check the Discovery Program page at **nssdc.gsfc.nasa.gov/planetary/discovery.html.** Another good source is the mission index page at NASA's Office of Space Science, which administers the Discovery Program. This page, at **www.hq.nasa.gov/office/oss/missions/index.htm**, is highly useful as it provides links to many missions, current and past, and not just those going to comets.

For a detailed look at NASA's "roadmap" for exploring the solar system over the next decade and a half, visit **sse.jpl.nasa.gov/roadmap/**. The site has links for downloading the PDF-format document, and for picking up a (free) copy of Adobe Acrobat Reader, which is necessary to open and read PDF files.

Comet resources in print

Rendezvous in Space, by John C. Brandt and Robert D. Chapman (W.H. Freeman, 1992). This book covers the modern science of comets for the interested lay reader. Although published before Hyakutake and Hale-Bopp, it gives a solid understanding of how comets work.

The Mystery of Comets, by Fred L. Whipple (Smithsonian Institution Press, 1985). Whipple figured out the physical nature of comet nuclei some 50 years ago. This pleasantly personal account of the whole subject is written at a more basic level than the foregoing title.

Comet, by Carl Sagan and Ann Druyan (Random House, 1985, 1997). This was first published in 1985 to catch the Comet Halley wave and reissued in paperback in 1997 for Hale-Bopp. The 1985 edition contains lots of colorful images of many Great Comets, but unfortunately the 1997 edition omits most of the pictures. Search in used book stores (or a library) for the original edition.

Fire in the Sky, by Roberta J. M. Olson and Jay M. Pasachoff (Cambridge University Press, 1998). This work, a collaboration by an art historian and an astronomer, looks at the depictions of comets made by British artists in the nineteenth century. It contains illustrations and color plates of many works of art showing such Great Comets as Donati's (1858), Halley (1835), and the Great Comet of 1811.

Comets, Popular Culture, and the Birth of Modern Cosmology, by Sara Schechner Genuth (Princeton University Press, 1997). Schechner Genuth has made a close examination of popular broadsides and scientific writings from the sixteenth and seventeenth centuries. She argues that political considerations led Newton, Halley, and their contemporaries to suppress the idea that comets colliding with Earth could cause catastrophes.

Comets, by Donald K. Yeomans (John Wiley & Sons, 1991). The subtitle calls this a chronological history of observation, science, myth, and folklore, which definitely fits. Nearly 500 pages long and organized more or less historically, it's an ideal source for looking up data on individual Great Comets, ordinary comets, or finding out how they work.

Comets, a Descriptive Catalog, by Gary W. Kronk (Enslow Publishers, 1984). Covering all comets discovered between 371 BC and AD 1982, this is a massive compendium of data on how a given comet was discovered, where in the sky it traveled, and what the apparition looked like. Not a resource you would sit down to read continuously, but rather one to dip into. (A new and much expanded edition, in several volumes, will be published under the title *Cometography* by Cambridge University Press starting in 1999.)

An excellent source for Greek and Roman writings on comets is the series of volumes in the Loeb Classical Library (Harvard University Press). These handy little books feature the original Greek or Latin text on the left page, a good English translation on the right, and detailed introductions. I used Loeb translations for the quotations from ancient writers in Chapter 6. See Aristotle (*Meteorologica*; trans. H.D.P. Lee); Manilius (*Astronomica*; trans. G.P. Goold), Seneca (*Naturales Quaestiones*; trans. T.H. Corcoran), and Pliny (*Natural History*; trans. H. Rackham, W.H.S. Jones, and D.E. Eichholz).

Comet Hale-Bopp (1997)

Comet Hale-Bopp: Find and Enjoy the Great Comet, by Robert Burnham (Cambridge University Press, 1997).

Everybody's Comet; a Layman's Guide to Comet Hale-Bopp, by Alan Hale (High Lonesome Books, 1996).

The Comet Hale-Bopp Book, by Thomas Hockey (ATL Press, 1996).

An Observer's Guide to Comet Hale-Bopp, by Don Machholz (MakeWood Products, 1996).

Comet of the Century; From Halley to Hale-Bopp, by Fred Schaaf (Copernicus/Springer-Verlag, 1997).

These books are chiefly observing guides and were written as Hale-Bopp was approaching the Sun, so they reflect mainly the results of Hyakutake and only the most preliminary findings about Hale-Bopp. Fred Schaaf's work is the most lengthy. It provides a lot of background about comets coupled to a narrative that relates his own personal observing history regarding comets.

Comet Halley (1910 and 1986)

The Halley literature is vast. Among the more interesting and useful items are the following.

Comet Fever, by Donald Gropman (Fireside Books, 1985); *Halley's Comet, 1910: Fire in the Sky*, by Jerred Metz (Singing Bone Press, 1985). These two books deal entirely with the 1910 Halleymania and both are a delight to read. Most remarkably, they don't overlap much.

Comets in the Post-Halley Era, edited by Ray L. Newburn, Marcia M. Neugebauer, and Jürgen Rahe (Kluwer, 1991). This two-volume compilation of scientific papers summarizies the knowledge gained by the apparition of Comet Halley and puts it into context. Not easy reading to be sure, but worth skimming in a university library for the picture it draws of a worldwide effort to study the most famous Great Comet of all.

Fire and Ice, by Roberta J. M. Olson (Walker & Company, 1985). A book by an art historian that traces the ways comets (especially Halley's) have been depicted by great artists over the centuries. Lots of illustrations and interesting perspectives from a scholarly field outside astronomy.

Mankind's Comet, by Guy Ottewell and Fred Schaaf (Astronomical Workshop, 1985). A superb work that shows maps and charts of all the known apparitions of Comet Halley starting in 1404 BC, and much other comet lore besides. A great book unlike any other and well worth seeking out.

Impact hazards

Comets: Creators and Destroyers, by David H. Levy (Touchstone/Simon & Schuster, 1998). An absorbing look at comets both as bringers of life's raw materials and as hazardous objects. Written for the lay reader by one of the co-discoverers of Comet Shoemaker-Levy 9.

Rain of Iron and Ice, by John S. Lewis (Addison-Wesley, 1996). An outstanding summary of the dangers posed by the impacts of comets and asteroids.

Impact! by Gerrit Verschuur (Oxford University Press, 1996). An engaging popular work looking at the risks of impacts and the role they have played in geological history.

Hazards Due to Comets and Asteroids, edited by Tom Gehrels (University of Arizona Press, 1994). A 1,300-page tome compiled by scientists for scientists, but this contains the most up-to-date technical survey of the threat and how to deal with it.

The Great Comet Crash, edited by John R. Spencer and Jacqueline Mitton (Cambridge University Press, 1995). A anthology of short essays by scientists who examine all aspects of Comet Shoemaker-Levy 9 — from discovery to its fateful meeting with Jupiter. Has loads of images and a text aimed at the interested lay reader.

Index

(Page numbers in **bold face** refer to images. Asteroids are listed in alphabetical order, followed by unnamed objects; comet designations follow the current IAU system.)

Alvarez, Luis, 196
aphelion, defined, 37
Armageddon (movie), 205, 208
asteroids
  9969 Braille, 152
  2060 Chiron, 19, 43, 53, 74–75
  5335 Damocles, 75
  4979 Otawara, 158
  3200 Phaethon, 25, 75–76
  140 Siwa, 158
  4015 Wilson-Harrington, 75, 152–153, 157
  1996 PW, 75
  1997 XF11, 208–**210**, 211
  1999 AN10, 211
  near-Earth (*see* near-Earth objects)
astronomical unit defined, 25–26
Australian aborigine legends, 213

Barnard, Edward Emerson, 184
Bessel, Friedrich Wilhelm, 12, 17
Beta Pictoris (star), **35**, 42
Biermann, Ludwig, 17, 29
Bigelow Sky Survey search program, 205
Bopp, Thomas, 100–101, 109, 134
Brahe, Tycho, 8
Bryant, Ted, 213
Burnham, Sherburne W., 183

Centaurs, 43–44, 74–76
Charon (moon of Pluto), 46
Chiron (*see* asteroids)
Comet Arend-Roland (C/1956 R1), **176**, 188
Comet d'Arrest (6P), 157
Comet Austin (C/1989 X1), 73
Comet Bennett (C/1969 Y1), 53, 61, 139
Comet Biela (3D), 15
Comet Borrelly (19P), **144**, 153
Comet Donati (C/1858 L1), 15, **16**, 69, 183, 222
Comet Encke (2P), 11, 17, 49, 139, **148**, 157, 203
Comet Giacobini-Zinner (21P), 142, 143, 144
Comet Grigg-Skjellerup (26P), 149
Comet Hale-Bopp (C/1995 O1), **cover**, **ii**, **10**, **18**, 21, 27, 30, **32**, 34, 35, **45**, **52**, **56**, 57, **59**, 62, 63, 69, 72, 73, 100–135, 149, 151, 157, 183, 190–**193**, **214**, 219, 221
  discovery, 100–101
  nucleus, 19, 25, 53, 101, 113–115, 120
  orbit, 103, **105**–109
  scientific findings, 131–134
  visibility, 109–131
Comet Halley (1P), 11, 17, 19, **20**, 21, 22, 34, 48–50, 53, 57, 70–72, 109, 123, **138**, 142, 143–149, 151, 154, **166–167**, **170**, 179, 183, **185**, 189, 221–222
  apparition in 1910, 183–188
  apparition in 1986, 189–190
  mass, 22–25
  nucleus, 19, **20**, 25, 53, 115, **139**, 146–149
  orbit, 34, 37, 72
Comet Hyakutake (C/1995 Y1), 77–82
Comet Hyakutake (C/1996 B2), **7**, **28**, 30, **54**, **58**, 69, **71**, 77, **78**–99, 116, 121, 123, 126, 129, 141, 149, **186**, 190, **193**, 219
  discovery, 77–82
  nucleus, 21, 96–99, 115, 141
  orbit, 72, **80**, 83–86
  scientific findings, 96–99
  visibility, 86–96
Comet Ikeya-Seki (C/1965 S1), vii, **60**, 63, 66, 69, **178**, 180, 189
Comet IRAS-Araki-Alcock (C/1983 H1), 49-50, 51, 53, 57, **67**, 69

Comet Kobayashi-Berger-Milon (C/1975 N1), 49
Comet Kohoutek (C/1973 E1), **68**, 72–73, 117, 139, 140–141, **180**, 189
Comet Levy (C/1990 K1), viii
Comet Levy-Rudenko (C/1984 V1), viii
Comet Lexell (D/1770 L1), 69
Comet Morehouse (C/1908 R1)
Comet Mrkos (C/1957 P1), **55**, 131, 188
Comet Schwassmann-Wachmann 3 (73P), 157, 158
Comet Seki-Lines (C/1962 C1), 49, **177**
Comet Shoemaker-Levy 9 (D/1993 F2), viii, 22, 197, 201, 203
Comet Swift-Tuttle (109P), 13
Comet Tago-Sato-Kosaka (C/1969 T1), 139
Comet Tempel (C/1864 N1), 15
Comet Tempel 1 (9P), **159**, 160–162
Comet Tempel 2 (10P), 142
Comet Tempel-Tuttle (55P), 13, 14
Comet West (C/1976 V1), 22, **23**, 49, 61, **64–65**, 66, 70, 73, **182**, 189
Comet Wild 2 (81P), 153–155
Comet Wilson-Harrington (107P), 75, 152–153, 157
Comet Wirtanen (46P), **24**, 25, 53, **154**, 157, 158, 160
comets
activity, 26, 57–61, 99, 113, 115, 121, 131
anti-tail, 130
Aristotle's views, 6, 169
asteroid-like, 74–76
Chinese views, 29, 177–180
coma, 25, 27–31, 53–4, 61, 101, 102, 113, 117, 120–121, 126, 129–130, 139, 141, 145–148, 154–155
composition, 14, 17, 63, 73, 96–98, 101, 109, 130, 131–134, 141, 143, 145–146, 155, 159, 161, 184
designations, 47–50
disconnection event, 31, **90–91**
dust, 15, 19, 26–27, 53, 57–61, 96–98, 101, 102, 113, 126, 131–134, 145–148, 153, 154–155, 157, 159–160
dust tail, 30–31, 69–70, 95, 117, 120–121, 123, 126, 129, 141
erosion of nucleus, 25, 131
and extinctions, 38–39, 41 (*see also* impacts from comets and asteroids)
fates of, 39–41, 43, 109
formation, 31–34, 41–42, 106–109
gas tail, 14, 30–31, 61, 69–70, 89, 117, 123, 126, 141, 142, 145, 153, 184–187
Great Comet criteria, 51
how named, 49
hydrogen halo, 29, 127, 139, 145
ice, 17, 25–27, 33–34, 96, 101, 102, 120
interior of nucleus, 21, **22**
and the Internet, 158, 188, 193–194, 218–221
ion tail (*see* comets, gas tail)
jets, 17, 29, 57–61, 99, 113, 115, 121, 131, 146–149, 157, 161
Jupiter family, 35–37, 161
Kreutz family, 63–66, 180
long period, 34–35
Manilius' views, 169
mass of nucleus, 23, 25
Medieval views, 8, 70, 175–177
new discoveries, 218–219
non-gravitational forces, 12, 17
nucleus, 12, 17, 19, **20**, 21, 53–61, 89, 98, 101, 120, 149, 153, 157, 160–161
orbits, 34–47, 72, **80**, 83–86, 103, **105**–109
and origin of life, ix, 43
plasma tail (*see* comets, gas tail)
Pliny the Elder's views, 174–175
and politics, 181, 195, 222
popular views, 8, 70, 164–194
Ptolemy 's views, 175
radar studies of, 98–99
rotation, 19, 57, 113, 160
Seneca's views, 6, 169–174
short period, 34–7, 157
showers of, 38
sublimation, 25–26
sungrazers, 62–**66**
surface materials and properties, 19, 21, 57–61, 159
X-rays from, **87**, 98
Copernicus, Nicholas, 8, 9

Danielson, Ed, 50
Deep Impact (movie), 57, 160, 205, 208
Deslandres, Henri, 184
Discovery-class missions, 151–152, 221
Donati, Giovanni Battista, 15

Edgar Wilson Prize award for comet discoveries, 50
Edgeworth, Kenneth, 42
Edgeworth-Kuiper Belt (*see* Kuiper Belt)
Edward VII, King of England, 183
Encke, Johann, 11, 48
Epsilon Eridani (star), 42

Flammarion, Camille, 187

Galilei, Galileo, 8, 9
Galileo spacecraft, 197
Gliese 710 (star), 38
Great Comet defined, 51
Great Comet of 1531, 11
Great Comet of 1556, **13**, 47
Great Comet of 1577, 8, 62
Great Comet of 1607, 11
Great Comet of 1618, 8, 61
Great Comet of 1680, 9, 62
Great Comet of 1682, 11
Great Comet of 1811, 51, 53, 62, 69, 182, 222
Great Comet of 1858 (*see* Comet Donati)
Great Comet of 1861, 70, 183
Great Comet of 1882, 62, 65–66, 183
Great January Comet of 1910, 62, 70, 71–72, 184
Great March Comet of 1843, 62, 63, 65, **168**, 183

Hale, Alan, 100–101, 109, 134
Halley, Edmond, 9, 11, 47, 48, 181, 195
Heaven's Gate religious group, 191–**192**
Hughes, David, 51, 70, 72
Hyakutake, Yuji, 77–82, 86

impacts from comets and asteroids, 195–217
  effects, 197–200
  end of the Cretaceous impact, 196, 197–200
  montioring from orbit, 212–215
  risks, 215–216
  tsunamis, 199, 213
  Tunguska impact in Siberia, 200–203, 212
InfraRed Astronomy Satellite (IRAS), 49, 75
International Astronomical Union, 48–49
International Comet Quarterly, 219
International Halley Watch, 158
International Ultraviolet Explorer (IUE), 141
Jewitt, David, 44, 50

Kepler, Johann, 8–9, 29
Kirkwood, Daniel, 13
Kowal, Charles, 74
Kreutz, Heinrich, 63
Kuiper, Gerard, 17, 41,
Kuiper Belt, **36**, 42–47, 85, 189

LINEAR search program, 207
LONEOS search program, 205
Luu, Jane, 44

Mariner 10, 141
Marsden, Brian, 50, 64–65, 208–210
Méchain, Pierre-François, 48
Messier, Charles, 48
Meteor Crater, Arizona, **202**, 203
meteor showers, 12, 204
  Geminid, 25, 76
  Leonid, 12–**14**, 15
  Perseid, 12, 13
Milani, Andrea, 211
missions to comets
  Comet Nucleus Tour (Contour), **147**, 155–158, 220
  Deep Impact, **156**, 160–162, 220
  Deep Space 1, 75, **143**, 152–153, 220
  Giotto, 19, 22, **137**, 147–149, 151, 154, 157
  International Comet Explorer (ICE), 142–144
  Pluto-Kuiper Express, **161**, 162–163, 220
  Rosetta, 24, 25, **150**, **153**, 157, 158–160, 220
  Sakigake, 144–146
  Stardust, **145**, 153–5, 156, 220
  Suisei, 144–146
  Vega 1 and 2, 146–147
Morrison, David, 203, 207

NASA finding for search programs, 205
NASA Near-Earth Object program, 207
Near-Earth Asteroid Rendezvous (NEAR) mission, 157
Near-Earth Asteroid Tracking (NEAT) search program, 205
Near-Earth Objects (comets and asteroids), 203–204, 219

1997 XF11, 208–**210**, 211
1999 AN10, 211
close passages, 208, 219
search programs, 205–207
threat, 204
Nemesis, 38–39
Newton, Hubert, 13
Newton, Isaac, 9, 11, 12, 47, 181, 195

Olbers, Heinrich Wilhelm, 13
Oort, Jan, 17, 37, 41
Oort Cloud, 34, 37–47, 64, 73, 83, 86, 134, 189
Orbiting Astronomical Observatory-2 (OAO-2), 139
Orbiting Astronomical Observatory-3 (OAO-3), 140
Orbiting Geophysical Observatory-5 (OGO-5), 139

perihelion defined, 39
Pioneer Venus Orbiter, 141–142
Planet X, 39
Pluto, 46–47, 162–163

Shoemaker, Eugene, 201, 205
Shramek, Chuck, 190–191
Skylab, 139–140
Solar and Heliospheric Observatory (SOHO) satellite, 50, 63, **66**, 95, **97**, 127, 220
Solar Max satellite, 62
solar system
formation, 31–34, 41
migration of giant planets, 44
solar wind, 30, 141, 143, 145, 163
Solwind (instrument), 62
Spaceguard Survey, 203–205, 207, 216, 219
costs, 203, 205
Spacewatch search program, 205, 207, 208

Torino scale for impact hazards, 211, 219
transneptunian objects, 44–47, 218–219
1992 QB1, 44–46
1996 TL66, 46
Triton (moon of Neptune), 46
tsunamis from impacts, 199
Tunguska event (*see* impacts from comets and asteroids),

Ulysses Comet Watch, 163, 220
Ulysses solar observation satellite, 163

Whipple, Fred, 17
Wolf, Max, 183

Yeomans, Donald, 51, 70, 72, 207